MADE IN ENGLAND

First published in 1939, Dorothy Hartley's *Made in England* is an unique and fascinating guide to some of the age-old country jobs and skills being practised in England today, from carpentry and joinery to thatching and hearthstoning, from the job of a smithy, to spinning, carding, dyeing and weaving.

Dorothy Hartley also includes a section on sea coast and riverside work, and succeeds in portraying the often unsung accomplishments of traditional England, from Sussex to the West Country, Yorkshire to Cornwall.

'A fascinating and beautiful book' *Daily Telegraph*

Also published in the National Trust Classics

Earls of Creation by James Lees-Milne
Ghastly Good Taste by John Betjeman
The Rule of Taste by John Steegman
The Wild Garden by William Robinson
Introduced by Richard Mabey
Felbrigg, the Story of a House by R. W. Ketton-Cremer
Introduced by Wilhelmine Harrod
In a Gloucestershire Garden by Canon Ellacombe
Introduced by Rosemary Verey
Bath by Edith Sitwell
The Housekeeping Book of Susanna Whatman
Introduced by Christina Hardyment

The cover illustration shows 'Outside the Blacksmith's'
by William Fisher

MADE
IN ENGLAND

Written and illustrated by

DOROTHY HARTLEY

CENTURY
LONDON MELBOURNE AUCKLAND JOHANNESBURG
in association with the National Trust

First published in 1939 by Methuen & Co. Ltd

© Dorothy Hartley 1939

This edition first published in 1987 by Century, an imprint of Century Hutchinson Ltd, in association with the National Trust for Places of Historic Interest or Natural Beauty, 36 Queen Anne's Gate, London SW1H 9AS

Century Hutchinson Ltd, Brookmount House, 62–65 Chandos Place, London WC2N 4NW

Century Hutchinson Australia Pty Ltd
PO Box 496, 16–22 Church Street, Hawthorn, Victoria 3122, Australia

Century Hutchinson New Zealand Limited
PO Box 40–086, Glenfield, Auckland 10, New Zealand

Century Hutchinson South Africa (Pty) Ltd
PO Box 337, Berglvei, 2012 South Africa

ISBN 0 7126 1705 7

Published in association with the National Trust, this series is devoted to reprinting books on artistic, architectural, social and cultural heritage of Britain. The imprint will cover buildings and monuments, arts and crafts, gardening and landscape in a variety of literary forms, including histories, memoirs, biographies and letters.

The Century Classics also includes the Travellers, Seafarers and Lives and Letters series.

Printed in Great Britain by
Richard Clay Ltd, Bungay, Suffolk

CONTENTS

IRISH HAY STACK — AFTER
THE CREATURES WOULD BE
GETTING AT IT

PREFACE

ALL COUNTRYMEN are not farmers or land workers. They have many other skilled occupations of which the average townspeople know little or nothing, such as thatching, weaving, making saddlery, work in the smithies, in the woods and quarries, and in the brickfields and potteries. This book describes some of these age-old country jobs and tries to explain the skill put into them. There are so many still being carried on that even to make a representative selection has been difficult.

I did not wish to include anything that might be called ' olde ', so have left out some ancient crafts of real interest which are likely to have been completely killed by the time this book is published. In addition, the intensely interesting domestic work and all the jobs done on coast and sea have had to be omitted from this volume for lack of space.[1]

All the work described here is alive to-day and most of the descriptions have been checked over by the workpeople themselves. I can guarantee that these things are still made in Britain, and that the drawings and photographs are taken from those actually in use by country workmen.

But what is done by one man in one way in one place may be done quite differently by another elsewhere. Therefore, before considering any apparent inaccuracy, the reader should realize how very widely usages differ in this small island. In every section has some expert gone over the description of his own craft, and we have removed anything that could not pass as ' general ' usage. (The only facts I cannot vouch for in this book are the Government statistics.) The very difficulty in finding any census of these country workers, or any official appreciation of their numbers, proves how unknown and neglected they are, and by a Government to whom they should be of the utmost interest, since they are

[1] See note on page 245.

the very root stock of our people, and little able to care for themselves.

Much of the work must seem very elementary to those who know the country, but this book is written for those who do not know it, because understanding makes appreciation and I want everyone to appreciate the work done by country people. Not for its commercial value, but because the work is done by independent people, and the character of these few independent people is as strong as the goods they make. It is these country men and women themselves who are of the utmost value to England.

Our large towns are no longer representative of this old English stock. The big commercial enterprises are mainly concerned with making money, and care too much for uniformity in their workers. It is usually in small places now that one can find the individual workman who deeply enjoys his work and puts it first in his life's interest. If we do not realize the value of these workers and appreciate and protect the human qualities they develop, we shall, later, have no standard by which to depreciate the mechanical robot resulting from the pressure of departmentalized industries. For it is not by the great works of a few, but by the life and work of the whole nation that its vitality is developed, and on this physical and mental growth does the people's life and work depend. A normal healthy man desires to work. The interest in his work is the growing point of a man's life, and once this growing point is broken, his driving power but presses against a blunted faculty. No artificial diversions or 'leisure' occupations substituted on the recreational plane of his existence will compensate him. It is as fundamentally wrong for a man only to work to live, as to live only for play.

By specializing work you can increase output, and give the worker more free time, but is he able to do anything new and vital in that free time if you have stunted his initiative, his creative faculty? Mass production and specialization in themselves need not destroy vitality, for where co-ordination and interest exist, vitality can persist. But the whole trend of factory industrialization to-day is towards a few clever minds directing well-drilled obedient masses, thus there is little appreciation of the individual country worker who has always been obliged to think for himself.

Therefore do not belittle my earnest descriptions of small workshops and little jobs, for perhaps if you study a job in its fundamental simplicity, you will more easily understand its growth and see what the individual worker has lost or gained by its commercial development.

But apart from all these abstract conjectures, this book is a plain record of country work, written as simply as possible. Half the work has been done sitting on ladders or tool boxes. Some of the sections have been re-done half a dozen times, for the first draft would be talked over by the workmen, who would make small changes, and then I would write it out again; next somebody else would be dubious about some point, and want to add some detail. The workpeople took all corrections very seriously and went to endless trouble to see that everything was ' properly put down '. So that by the time an entire quarry or workshop felt they had ' got it fixed ', I'd have to go home and write it all out again, and yet again, till—

' Of course there's a lot more to it, but what you've now got's right, you've got nothing there I'd need you to alter; it can stand now,' would be the belated benediction of my experts, and the reader can reach the final word with no more relief than did the browbeaten author when at last permitted to write it down.

The drawings were usually made on the spot, the proportions roughly measured up by hand (palm stretch 7 inches, pivoting 14 inches). The photographs were taken while the workman was actually doing the job, so your indulgence is asked for many taken under very difficult conditions.

I should like to thank all the people who have taught me, especially those who so painstakingly corrected notes and drawings and puzzled out the local names for the tools. Also the tool manufacturers and others who helped me to correlate these local names with the trade terms (where possible).

Also my publishers for their patience: without their

encouragement I should have despaired of ever getting the debris of ten years' research work cut down into this one small volume.

It is full of faults, but I do hope, between us, we have managed to give some real pictures of country life and work.

Vron
1931–1938

I am glad of the opportunity to make some corrections and additions, and to explain to my many kind reviewers that the original MS. had to be cut down so drastically, it went to press under my protest that there was not enough left to be worth printing.

I thoroughly sympathize with the reviewer who wished I had not wasted space describing landscape and weather at the expense of technical detail, though the weather is a technical detail to country workers, and a mountain is not a landscape to a shepherd, it is a sheep walk.

Finally, I should like to thank Mr. Frederic Littman, who rescued the original book from exasperated oblivion and carefully prepared it for press, with all the proof reading : and how awful proofs are after the author and the artist and all the workers have all made all their corrections—only he knows.

D. H.

Vron
1939

PREFACE TO THIRD EDITION

War and Peace have passed in ten years over the workers in this book. Some have been killed in a Malay jungle, or African desert. The old man (page 176) died gently in his cottage by the kilns, his smooth worn hands are quiet, and his wheel still.

The young farmer (page 65) carting green rushes down the bare hillside was invaded by half a London slum and a factory (he remained unmoved, so did they). The straw bed (page 74) was issued to a Corps of Signals, bedded on the stone floors of a Western castle, and ' they slept warm, with " Wyspes drawen oute at fete and syde ".' The use of dyes has developed, since Elizabeth's seamen were told to ' learne all manner of dies ' worthy of our excellent cloth. Shortage of petrol has caused horse collar repairs ; lack of matches flint cutting ; ' Wood ' takes a deeper meaning, now England is being reafforested. The same smoke from the charcoal burner's fire drifts down a thousand years, and the Rodmen of the Severn face a crisis that was old when King Alfred was new.

The leather muffle for the passing bell has been wet and dried again for a few old friends, but their grandsons use their tools, and the wind and the rain shape their tally to the same pattern of Peace. The Peace of Craftsmen—who *' are not found where parables are spoken, but they will maintain the Peace of the World—and all their desire is in the work of their craft.'*

VRON

1950

PREFACE TO FOURTH EDITION

I am glad of this reprint. The first edition came out during the exhibition on the South Bank in 1939. The bank was packed with pavilions, all exhibiting English Enterprise, but one remote pavilion called 'The Lion and the Unicorn' portrayed the English temperament. It was, I believe, the inspiration of one Peter Stukley (of family as old as England), and this pavilion completely puzzled foreigners.

Over the entrance, where flew the white Doves of Peace, flew also the colliers' racing pigeons. Alice in Wonderland knelt on a mantelpiece passing through into the Looking Glass world, where the dear old Knight, with his beehive and gentle smile, toppled regularly, head first, from his charger —the charger with fierce spiked hooves. There was a medley of sport—cricket bats, tennis rackets, golf clubs, longbows, footballs and boxing gloves, ponderous bowls, fishing rods and a pint pot by a dart board. . .

Alone, above, majestic and impressive, hung the great Sword of Justice and on a green baize tabletop alongside a dog collar and a pair of gardening gloves was the first edition of this book. Oh, the thump of joy to my young heart! My people were here ; they would be coming up, by excursion trains, to St Pancras, Paddington and old Doric Euston. They would find themselves here, and know their work was behind all the show outside. All the 'Enterprises' depended on the English working man. This first edition was their work.

The war left them, but took their sons, the boys they taught their craft to, and life lost its meaning when they were alone. One after another the old workshops closed down. Old tools went to the scrap-yard, for the hands that used them were gone. This is the great change today. These men took pleasure in their work. The boys decided what they wanted to be and fought to get training for it. Now they ask first, 'will it pay'? Have we lost the singlemindedness of direct drive? Always there will be opal-minded ones, whose golden mist reflects the ever-changing flame of life, but here, in this book, live the solid men who trusted in their hands and were wise in their work. *'They shall not be found where parables are spoken. But they will maintain the state of the world, and all their desire is in the work of their craft.'*

VRON   D. H.
1973

" All these trust to their hands :
    and everyone is wise in his work.
    Without these cannot a city be inhabited. . . .
    They shall not be sought for in publick counsel,
    nor sit high in the congregation :
    they shall not sit on the judges' seat,
    nor understand the sentence of judgment :
    they cannot declare justice and judgment ;
    and they shall not be found where parables are spoken.
    *But they will maintain the state of the world,
    and all their desire is in the work of their craft.*"

                                                                    ECCLESIASTICUS

CHAPTER ONE : WOOD

§ 1. *The Trees and their Uses*

Now that a tree can be felled, dried, and sliced up as conveniently as a sandwich loaf, the proportions of the cheap, mass-produced, white-wood furniture on the market are governed by the dimensions of the standard plank and batten.

This type of mass-produced, jig-saw work makes the transition from wood to synthetic substitutes unremarkable. There is no reason, apart from considerations of weight, cost, and surface finish, why wood-pulp board, asbestos, chromium, glass—almost any material—should not be substituted for ' wood '. There is no ' treen ' in the design of such furniture, they are not made of wood for anything but convenience.

By contrast, the most modern bent-plywood goods (from Norway, Sweden, and Finland, our godfathers in timber), show an understanding of wood and an appreciation of its qualities. They are as fine in their new light design as the heavy timber and carved hammer beam roofs of the old centuries ; thus proving yet again that good design depends neither on period, nor fashion, but on appreciative joy in the qualities of the medium.

The modern scientific skill that makes it possible for this light woodwork to retain its springing qualities, its grain and surface variety, and in its new form hold strongly all the essential goodness of wood—*that* is craftsmanship. But the chopping of wood into commercially convenient lumps for misusage by the jerry builder is an evil waste of one of the most beautiful materials on earth.

This comment on material and design is made because —to the country user—wood is still ' treen ' and the countryman's work, therefore, begins with his appreciation of trees, and their varied qualities *as material* for his job.

Wood is not inanimate. A tree, standing, is a mass of

conflicting stresses and strains. (If it was not, it could not stand.) A growing tree, wrongly felled, may split with a loud report, cracking open its entire length simply through wrongly released tension at the base. The relation between the outer and inner rings of the tree trunk is one of constantly varying pressure, and timber is capable of movement and development for years after it is cut.

Warping, shrinking, or ' shaking '—' shakes ' are the technical term for the cracks resultant on shrinking—swelling, are facts differing in various timbers as physical distinctions

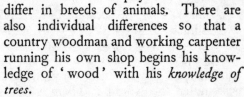

differ in breeds of animals. There are also individual differences so that a country woodman and working carpenter running his own shop begins his knowledge of ' wood ' with his *knowledge of trees*.

Forestry work is now a definite Government department, the varied branches of research, the draining of land, the planning and planting, thinning and felling of plantations, coppice, and park trees, being all parts of a huge Government industry, employing some 12,000 men. In addition about 1,700 are given as agricultural and forestry pupils ' not at colleges ', so presumably working.

Many trees now being planted are recent importations, but only the following are, from long habitation, usually considered ' English timber '.

OAK. There are two kinds, *Pedunculata*, the leaves having no stalk, but the acorns swinging, and *Sessiliflora*, in which the leaves are stalked and the acorns sessile. The *Sessiliflora*, resembling chestnut, is sometimes considered inferior. Both oaks vary much according to the soil, clay soil making a tough wood, which may account for the old reputation of Essex coastal oaks for ships and of Mildenhall oak for beetles (i.e. mallets). Previous to the importation of pine, all heavy building and structural work was of oak. The ' timbering ' and old oak of English architecture is a study in itself. The beams are all adze and cloven work (not sawn); pillars and

2

posts (in barns and buildings) are used root upwards to prevent damp rising, and evenness and symmetry are got by pairing the two split halves of the same tree, or branch. (Hence the country saying: he's the 'splitting image'—an exact likeness.) The root stump of an oak is often used by blacksmiths for an anvil base (see p. 147), and the bark makes the finest tanning. Oak sawdust was the smoking medium for York hams, and oak galls made ink. Old oak cart-wheel spokes are often 'cut down' for use as ladder rungs. An interesting note in all medieval calendars is the autumn fattening of pigs by beating down the beechmast and the acorns in the woods where they herded. Acorns are supposed to make a very hard bacon, but the description of sides of bacon 'that lookes like strips of leather' does not indicate very fat pork.

ASH. A wood as tough as oak, which will not splinter, so that for wheelwrights and for the making of long agricultural implement handles and for all purposes, such as aeroplane construction, needing tough tensile strength, ash is chosen. Grown as coppice (small timber), ash is used for stakes and poles, cart shafts, lorry bodies, chairs, tennis racquets, hoops, and small blocks and tool handles. Stooled ash is cut-back ash, the second and third growth from the ash root. Naturally bent ash poles are sometimes split for special curved harvest ladders, which can be pushed down off a stack, and not break.

ELM. This is a curious wood, very good against continuous damp. Hence it is used for coffins, piles under river bridges, quays, and ships' timber below the water line. Cart shafts, when not of ash, are sometimes elm. Elm turns very smoothly, so is good for rollers, for the dead-eyes in a ship's rigging, and for bobbins of trawl nets; boards and table tops are often of elm, but not in dairies or where there are alternate wet and dry conditions which soon rot it. In some timber yards all elm trees are 'barked' for their better preservation, and then their gaunt, smooth, silver-grey trunks show up among the rough oak or crusty ash.

In early autumn old elm trees often drop their boughs suddenly, in full weight of leaf, so the country people say,

3

of old folk who fall ill and die so suddenly, ' the hale old goes as ellum boughs fall.'

BEECH. A white wood (sometimes greyish), which when new cuts like cheese. For some purposes, such as kitchen use, this whiteness is good, but for handles and the like, where a dark colour is better, beech will take a very even stain. Some beech is used for clog bottoms, and dairy utensils. Country turners who use beech often set their blocks in the smoke of the fire (the waste fire of bark and chips that smoulders near every camp), the theory being that the acid sap of the beech is thereby chemically changed and the wood rendered more waterproof. Antique beech bowls and platters, however, have usually gained their waterproof durability from repeated servings of hot mutton fat. The modern preservative stains take well on beech and much of the pleasant wood turnery now popular for table ware is pickled beech.

SYCAMORE. This wood is largely used for ' fancy ' turnery, giving a very close, clean white finish, and also for sink draining boards, as it looks well against white pottery, and scrubs well.

BOXWOOD. Though seldom found as large trees, boxwood is very close and fine, and fetches a good price from the mathematical instrument makers for all precision work. When carefully pickled and bleached in the dark, it is used for wood block engraving (which has been done across the *ends* of the grain, since Bewicks's time).

WILLOW. Some varieties of this very springy wood are essential for the making of cricket bats. Being waterproof, it is also used for boat paddles and propellers under water, and for water-wheel slats. It stands heat fairly well (so you can oil a cricket bat before the fire), and in old kitchens willow bowls were used for the dripping to ' set ' in. The coarser willow, iron-banded, often forms the brake blocks on rolling stock, and the country wheelwright likes a chunk for his wagon block. It is interesting that green willow rind in vinegar was an old cure for warts, while nowadays the extracted salycylic acid is painted on them.

POPLAR. This wood has many of the qualities of willow.

4

ALDER. A tree unfortunately becoming very rare in England. It grows to a fair size by water, and gave the most excellent, absolutely waterproof wood for clogs and butter boards. The scarlet withys bunched by streams in the North show where hundreds of alders have been cut—but no new ones planted. Alder was also the best fuel from which to make charcoal for gunpowder.

BIRCH. The mountain tree which, because it will grow in high altitudes, gets used for everything there, whether it is suitable for the purpose or not. In the north it was the aromatic birch that gave quality to Scotch whisky distilling, and the golden excellence of the old smoked haddies and herrings was due to this resinous aromatic wood being used. In the valleys in the south, bundles of the pliant twigs of beech and birch are used in the making of vinegar.

WILD CHERRY. This, like some other lovely woods, is very local in growth, and often seems in a curious way to follow Roman roads.

WALNUT. Up to the eighteenth century this was the cabinet maker's wood, and it is to-day valuable for furniture if of good growth. It was (and is still) also used for gun stocks, and so during the Napoleonic Wars the price went up, and the trees came down, till England was nearly cleared of walnut trees.

They say,

> Beef steak, wife and walnut tree
> The more you beat 'em, better they be.

So country lads in spring go up and hack about the walnut tree with a hedge bill, till the sap runs freely. Some say that the sap attracts insects to 'set' the nuts; some, that the shaking distributes the pollen; and others that the bleeding sap induces fertility. I only know that, with our two walnut trees, during twelve years, we only had a phenomenally large crop the three springs I beat them. Old men affirm that the most 'beaten' trees have the hardest wood for gun stocks and the best market for 'figured' furniture. (Elm is sometimes 'Bole-trimmed' for the same reason.)

HORNBEAM. A very localized tree that was sought after

5

by millers for the cogs of their wooden wheels, apple wood being their second choice.

FIR AND PINE. These were grown on waste land, as private speculation for pit props, building construction, scaffolding and masts, till the imported timber made it uneconomic to replant. The enormous plantations of fir, pine, and larch now growing are chiefly the result of the new Forestry Department. Roughly 120 trees of Douglas fir go to the acre and 200 of larch. In some cases special seed is imported and grown on the nursery slopes, and the seed may come from as far as the American Pacific coast or from Japan.

There are also plantations of special trees in selected districts, such as wind-breaks by the sea, fox coverts in hunting counties, the coppice work of Kent for hops, and of Hampshire for the sheep hurdles and for young fruit tree stakes, and the smaller hop poles (though poles are now generally imported timber). The hedges usually belong to the land, to the farmer and agriculturalist, while 'coppice' (not timber in the heavy sense), is often specially grown for some definite market. These special plantations and coppices are all made by the countryman.

§ 2. 'Sawn' v. 'Split'

The very vital point for an elementary study of things made from wood is the distinct difference between 'split or riven timber' and 'sawn wood'. Few people realize this difference. Nowadays, when the word 'timber' in ordinary commercial use means ¾ inch, 1 inch, or 1½ inch deal, in ready made widths, any length cut to order, wood *as a texture with grain* is ignored by the average user. They order it ready made as 'shelfing', 'matching', 'quartering', 'battens', or 'boards', and these are all sawn. For a great many things this does not matter, but for some purposes sawn wood is utterly unsuitable.

As a small example, take a simple thing like a clothes peg; if it is turned out the wrong way of the grain, then the first time it is used on damp clothes, the peg will split and fall

6

apart. Or take a tent peg; you can cut a dozen from one slice of sawn timber, just as easily as you can cut a cake into a dozen wedges. But only two of these pegs may be of any use, because the twist of the grain will lie across the others and at the first blow their heads will come off. If their heads do not come off, their tails, coming across the fibre, will fray out like a bottle brush.

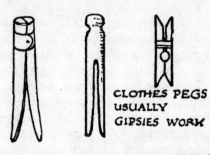

CLOTHES PEGS
USUALLY
GIPSIES WORK

Gates which are made of sawn timber look perhaps neater at first with their straight sawn lines, and, if the sawn timber has been well selected, may hold for an astonishingly long time, but, a bang on a wet day, and they will split. Any farm girl chopping sticks for the kitchen fire learns to utilize the cleaving properties of wood, and these properties *are used* to advantage by the 'cleft wood' worker.

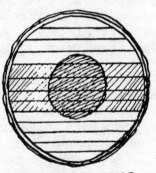

STRAIGHT SAWING

Timber which is allowed to follow its own grain in splitting and cleaving may afterwards be turned and finished to any shape required, even better than sawn wood, and it will have the advantage that it will not again split, neither will it soak up the rain, since the smooth lengths of the fibres are exposed, not the porous ends.

Cleft *v.* Sawn is often the whole subtle difference between good work, with well-chosen, suitable material, and a cheaper, inefficient substitute.

The long endurance of antique chairs and tables, even when finely turned, is due to their being made from naturally cleft timber, and for all outdoor purposes the cleft oak gate, the split pale fence and the riven peg more than repay those

7

who use them, for they outlast five times those made from sawn timber.

§ 3. *Timber Drags, or Drugs*

The large timber yards by sea and river docks usually deal with imported timber, ready cut, and are town premises employing townsfolk, but smaller timber yards up country deal in both imported and home grown timber.

In most cases the owner, or manager, travels and buys his timber abroad. Many office-bound townsmen would be surprised at the world-wide travels of some ' small country wood merchant '. The man who deals in English *local* timber only is usually either a specialist, or deals in a very small way, unless his yard happens to be the clearing house for a very well-wooded locality.

Some of these timber yards have their own horse teams and timber drags, motor haulage contractors undertaking the long distance work. But there are many compromises. Horse teams cannot go long distances, but are more mobile and convenient for local work. The most modern motor drag has convertible wheels with which it can climb through the woods like a cat, or, with its ' claws in ', roll smoothly along the high road.

The same motor drag also has winding gear, a derrick and chains; with such an outfit three or four workmen can fell huge trees and extract them from most difficult situations.

The horse team workers are no less cunning, and though their pulleys, and cross pulleys, and manœuvrings may be less spectacular, yet a proud wood team leading home a loaded drag is a fine country sight, and our lumber men, though few, are a good breed, who have brains and use them.

One small item in any load is the spring pole, that is the small sapling bent to hold the tension of the binding chain. Every timber drag uses it, yet I have never found any team that could give it a regular name. I asked an Austrian woodman what he would call it, but he was also at a loss, for apparently in no language is there a dictionary name for

8

this simple gadget.   It is 'the-springing-sapling-that-is-wedged-into-the-frame-of-the-timber-drag-and-serves-to-take-up-the-slack-of-the-binding-of-the-load",—*alias* 'spridger', 'bill springer', 'Nancy', 'the dancer'—any local nickname.

Loading timber from the drags on to trucks in the railway goods yard is again a specialized job.

The photograph (Illus. 2) shows a good example of long distance timber hauling, and was taken on a high road in North Wales. The whole job occupied some five days : the first day, spent in driving across from the east counties ; the second and third day, prospecting the wood and ' clearing out ' ; fourth day, felling and hauling, and either late that night or early on the fifth day, they would drive back with the heavy load.

The felling was on a dangerous slope above a road built out over a considerable ravine.   The timber was huge, and, working with great skill, and a cleverly used derrick, the three quiet men, all alone, had carried through their massive toil.

It was somehow funny to see them immediately afterwards wipe off the sweat, slip out neat little notebooks and white tape measures, and measure the tree, with all the neat accurate solicitude of a tailor measuring the waist line of a corpulent customer.   They worked delicately and accurately because their pay varied according to the cubic capacity of the timber measure, for timber is usually bought and hauled by the cubic mass, as piece work.

§ 4. *The Timber Merchant and his Yard*

The timber merchant is a dealer and trader rather than a manufacturer. His work is in the raw material, ' making ' baulks, blanks, or posts, though he may sometimes contract for squared and measured timber, thus touching the carpenter's end of the building trade, or for fencing fields, encroaching on the agriculturalist's work. But we have mentioned the complete difference between ' sawn ' and ' cleft ' or ' riven '

9

timber, and, speaking generally, the country man's experience prefers the enduring qualities of the latter.

The country timbermonger's skill is in cutting, seasoning, storing, dressing, and sizing his timber according to his knowledge of its character and the requirements of carpenter, joiner, wheelwright, building contractor, engineer, agri-

STACKING AND PILING TIMBER

culturalist, or miner. In addition to this knowledge of his commodity, he has generally much geographical knowledge (local and foreign) of its origin, so that in his job he may well be one of the best-informed of working countrymen.

The average timber yard will have its own derrick and employ men sorting and piling the raw timber, for timber

OFF-BARKING A TRUNK; AND USE FOR
STRIPS OF BARK-WOOD.

must be moved and stored so that it seasons evenly. In a large yard there will probably be a small-gauge railway and a donkey engine, to facilitate shunting about and hauling. The sheds for the cut timber must be open to the air, but safe from damp. Piled and stacked timber is 'built-up' in regular courses, for great damage can be done to timber by stacking it wrongly. There will be the yard office and work sheds, probably a lorry for delivery. They may, or not, have their own timber drag (Illus. 1).

As a rule the power is supplied by a steam engine, using a proportion of waste wood fuel. Sometimes water power is used in a sawmill. There will be band saws, and probably chamfering and planing tables and drills, and so on. Any yard dealing with heavy timber will have a rack bench and probably one of the new circular saws with replaceable teeth.

The boss or manager will often keep a horse to ride to the woods to see the men, and a small car to drive to town to see the customers.

Such a country timber yard may employ anything from five to fifty men, according to its size. Sometimes such yards are central depots for forest camps working at considerable distance. These camps may last perhaps a year, till the wood is used up, and while working, constitute a small self-contained community that usually vanishes on bicycles from Saturday afternoon to Monday morning, leaving drowsy horses and a somnolent caretaker to a pie and a kettle, with instructions to be sure to start the boiler up Sunday evening.

'Once a man has taken to timber, he's set,' the country people say. It is a hard outdoor life with, on the whole, plenty of variety in it.

It looks very peaceful, but the following questions from qualifying papers on 'wood' set at a timber merchants' examination, give some idea of the brain work required and the knowledge and wisdom of the 'simple country timber-monger'. (In all papers it is taken for granted nowadays that imported timbers are used.)

'*Explain the meaning of : scanfin, riftgrain, double wrack, cup shake.*

'*Describe a stack pile, cabinet pile, square and Bristol piles.* (This question is interesting, as many ports and quay yards have special forms of store).

'*What defects should be looked for in square-edged Honduras, hazel pine, steamed black walnut, hickory, American ash logs, quartered red oak ?*

'*In the special measurements for timber, convert 2,545 feet into* (a) *Pet. standards to 3rd decimal point;* (b) *Superficial feet of* $1\frac{3}{16}$ *inches thick.*

' *In plywood* (now stocked by the average country timber-monger) *what class is most suitable for : door panels, drawer bottoms, dust boards, and boxes for export, and why?*

' *What is meant by " wet cemented plywood " ? '*

' *What are stock sizes of Finnish birch, Russian birch, Norway pine, Oregon pine, Gaboon ? '*

There are also queries on forwarding a mixed parcel of standing ash and beech, and arranging for haulage and delivery.

Now—why should townsfolk generally suppose that a man needs less brains to make a living in the country ?

§ 5. *Workers in the Woods*

Differentiated from carpenters and bench-workers, or those who use sawn wood, are the number of wood-workers who work on the raw ' treen ' in the woods. Theirs is a craft by itself, and they must learn to select their own timber, travel afield to find it, and often (with the help of the local smith) make their own tools and appliances, as few of the commercial tools can be directly suited to their extremely varied type of work.

This trade of the woods has various names in different localities. In Buckinghamshire and the south, the workers are called ' bodgers ': probably a compromise between the old French ' bouger ', ' to move about ' and ' bodger ', a rough (not necessarily clumsy) block out, or patch. It is almost the equivalent word in wood, of the ' clout ' of the leather worker or cartwright. Both words have deteriorated in use, for nowadays ' a clout ' is a very rough patch indeed.

The workers, of course, move about following their timber, and to a certain extent their market. In Buckinghamshire they make rungs for the large chair industry, which still continues on the spot. The chair legs have mostly walked under cover into factories now, but many other bodgers still work out in the woods. One party, near Reading, have recently finished an order for several million tent pegs, and a smaller order for wooden brush back slats. The same type of worker in and around Carmarthenshire may specialize in wood bowls and spoons, and elsewhere in Wales or the North, using different timber, in clog bottoms.

In Kent, the workers in the coppice make palings and hop poles, while in other places you can find wood and coppice workers making gates and ladders (Illus. 5).

In some parts of Surrey and Sussex they used to make shingles, wooden tiles for roofing, and these are coming back into use now as the new wood preservatives make them more durable and less inflammable.

Where timber land is cleared a small circular saw is often used, and piles, fence stakes, perhaps gate-posts and firewood blocks are turned out rapidly on the spot from the freshly cut wood. They burn the bark and refuse, and the smoke from the chip, bark, and sawdust fire, floating up, has often led me to the spot long after the hum and squeal of the saw has died down for the night.

I saw a funny thing happen by one of these circular saws; it was in a Yorkshire wood, up the dales. The rotating saw hit an old strand of wire, which years ago had become grown over and buried into the tree. (Part of the overseer's job is to look over all timber, especially near fields or fences, for this danger). The spinning saw landed fairly on to the buried iron—there was a crash! The wire saw cage protected the workers, but a section of the broken saw spun upwards through the air, phist-wizz—to stick and hang quivering high up overhead in a tall tree just under a rook's nest. We all stood there, gaping, with our mouths wide open, heads back; and then in the sudden silence of the stopped engine, we saw an elderly black rook, with brood-stiffened legs, get up off that nest and waddle down the bough. She looked at that saw; she looked at it blamefully, she looked at it first with one eye, then with the other. . . . Then, with one firm 'pluck' she pulled it out and dropped it, watching it fall, slattingly, down, down, down, till it struck the earth. She watched it land. Then, slowly, she turned and waddled back to her nest, and settled down again, with her back turned at us.

We none of us spoke for ages. We just stood there in silence, looking up and feeling, somehow, awfully snubbed. I don't believe all the Board of Trade regulations and notices posted up all over the saw shed could have made us feel so rebuked as that rook.

Presently old Dave said, in a funny apologetic sort of

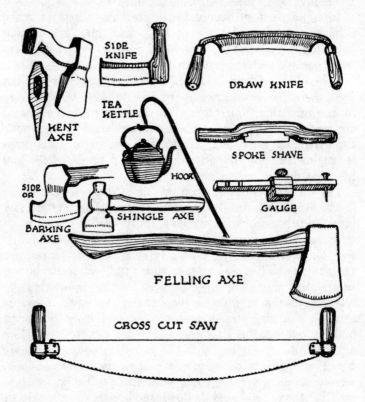

voice, ' It were an accident, missus,' and, like a spring released, we all began to laugh—and we fitted another saw.

The wood-workers' equipment varies with their job, and the distance from home. There is usually a roomy shed (built from the surrounding timber), often with a corrugated iron roof. On to the roof the workers throw all the bark strippings and chips, till it is thatched over thick. This is to stop the din of the rain (so that they can hear themselves think); also this thatch keeps the shed warmer in winter and cooler in the summer. As a rule the wood is shady, but if the hut is situated in a clearing, a couple of buckets of water thrown up on to the porous thatching on a hot day will cool down the shed if there is any breeze. In the shed are built, or set up, simple work benches. If there is any turning to be done, a pole lathe is contrived from a suitable sapling.

The next item is, of course, the rubbish fire with an iron hook for the tea kettle. And very pleasant it looks with the billy-cans all standing round, and the boss's teapot in the middle. Near the fire is the box for the boy to sit on, and keep the condensed milk cans in. I nearly counted the boy as part of the equipment, for there is always 'the boy' about a camp—he just comes, like a cat to a house.

And then there are tools, sacks, and water tubs, varying according to requirements of the job and the time they will be doing it.

§ 6. *A Beech Wood Camp Making Tent Pegs*

This camp was in the Buckinghamshire beech woods. I found it through noticing a chip of fresh-cut wood sticking in the mud on the foot-rest of a stile. It was a piece of white beech, near heart wood, so freshly cut that it was still damp; so I went in to the nearest beech wood and looked for the camp.

It was a beautiful spot; the wild cherry was in bloom, and wood violets were among the curled brown leaves.

The shed was commodious, holding three or four 'places'. The 'benches', solid blocks of cut beech (which in their turn would be cut up) were set towards the front of the shed in the best light. There were some pegs driven

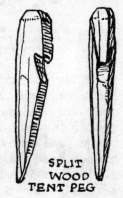

SPLIT
WOOD
TENT PEG

into the walls for the men's coats, some folded sacks for sitting on, a shelf for whetstones, an oil-can, waste, mugs, empty tobacco tins, the newspaper, the account book, the stub end of the pencil that would just write (and the stub end that wouldn't), and any odds and ends a man likes to have handy. There was a nail driven in for the time-keeper's watch. The earth floor was padded warm with shavings (Illus. 3).

To hold the water there was a hàd-been-white-enamelled Oxford hip bath.

(I don't suppose they'll ever bother to take that hip bath back; it will sit for ever now in the Buckingham beech

HIP BATH THAT TOOK TO
BUCKINGHAMSHIRE BEECHWOODS

woods, and the leaves will drift into it, and the rain will puddle, and the little birds will bathe in it, until the bottom gives out. And then the ferns will grow through it, the rim will drop off, robins will nest in the opulent ornamental handles, and finally the ghost of that Hip Bath will go flapping back, thin as the wraith of its vanished enamel, to the Pump Rooms of Bath and the upholstered hotels where hip baths belong. And the other hip bath ghosts, and the foot baths, and all the properly plated H's and C's will see the chips in its hair, and the ivy leaves upon its brow, and think how one respectable Hip Bath brother once

16

lived a wild life, and went to the bad in the Buckinghamshire beech woods.)

The selected beech trees are lopped and felled close to the ground with a cross-cut saw. This is monotonous work, but the wood cuts sweetly. Once down, the timber is sized up and measured off as it will cut best. In this camp, they were making tent pegs, so instead of any complicated calculations, they merely laid the standard tent pegs along the tree and marked with a rub of the saw where the cuts came. The slices of tree were the size and shape of Cheshire cheeses (the old Cheshires were white, before they took to using annato, a sort of colouring matter).

These round sections were then rolled to the shed and split. This splitting is the essential characteristic of good woodwork, as I explained on pp. 6 and 7.

The blocks are then collected and dumped by the side of the iron cutter: with half a dozen movements, the worker shaped the blocks on the fixed knife, and when he

THE FOX & THE TENT PEGS

had done a good pile sat down and finished them by hand with a draw knife. As the pegs were made they accumulated in a pile till he could see he had done enough for a stack.

The next proceeding was to take the pegs out and stack them. These turrets of white pegs looked like white honeycomb in the woods. The small wild things, who run away when the men come, presently grow used to the invasion, and, after the dog has gone home at night, they return (for wild things are very curious) and investigate. Once, by moonlight, I saw a fox go through. Perhaps he connected the white wood with hen pens, for I saw him stop and sniff at the pile and then he sat up springingly like a thin dog begging, and I thought he was going to put one black pad against the wood to steady himself and the pile would tumble over, . . . but he thought the better of it, dropped down, went and had a look at the hip bath, walked round the fire (I expect he'd found a tit-bit some time), and then went off. Later a hare went through, and a mole fetched some chips. I wondered why he wanted them? In the early dawn the birds came for crumbs. The men did not arrive till fairly late in the morning : most of them came on bicycles. They don't start woodwork till the dew is up.

When the consignment was finished and had seasoned long enough, the pegs were packed into sacks and sewn up for dispatch. About a week later than the appointed day (this was a government purchase), the lorry came round and crashed and wriggled its way through the wood (a little axe-work was wanted here), loaded up the pegs, signed up receipts and papers with the boss, and took them off (Illus. 4).

The shed stayed there all next winter, because it might as well be there as anywhere. Then it was moved ; one of the men used the timber for a garden shed, and they took the sheet of corrugated iron roof to a job elsewhere.

§ 7. *Wooden Spoons and Spurtles*

I found the wooden-spoon makers of Carmarthenshire after seeing their goods on sale in the market stalls at Cardigan : these spoons were very noticeable because of the individual shape.

The spoons, the various dairy utensils, and some queer shaped scoops which I subsequently found out to be churn laths, were obviously all the work of different craftsmen. (If you have a ' worker's eye ' you see this naturally, but you can always prove it to your satisfaction by measuring the tool-marks, for each worker of course uses a different tool.)

Through the retailer, I succeeded in tracing the maker of

the spoons to a small farmhouse beyond Llandilo. This district (to the west of Forest Fawr) is well wooded, pastoral country. Hence, probably, the prevalence of dairy utensils. The wood-worker, Mr. Jones, relied on getting his timber locally and employed odd haulage. When I was there he was working single-handed, though occasionally, if he had timber and orders waiting, he would find someone to help

POLE LATHE
13TH CENTURY

BENT SAPLING

CHUCK

TURNING A
WOODEN
BOWL

TREADLE

him. His work-shed was part of the outbuilding of a small farm. He had purchased a consignment of wood from where some lumber men had been cutting, and so it was already in short lengths. The pole lathe was used for some of the spoons, both in rounding the bowl and turning the handle, but in the majority of hooked spoons the handle was cut by hand, not turned. The pole lathe remains unchanged down the centuries. The rotating action is between a bent sapling and a treadle.

19

This worker had on hand a consignment of round-seated milking stools, three-legged. The usual milk stool has a seat squared on one side between two legs and a third leg behind. This is for convenience in tilting forward. The

MILKING STOOLS ARE MADE TO TILT FORWARD

round stools that were in use in this locality were as individual as the circular milk cans, which I noticed lower down the peninsula towards St. David's (where the Flemings settled).

Besides the common thick spoons of everyday usage, he had also at various times turned out some exquisitely delicate

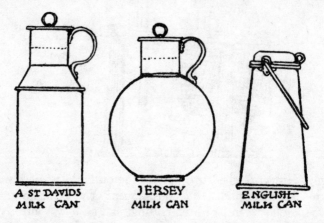

A ST DAVIDS MILK CAN   JERSEY MILK CAN   ENGLISH MILK CAN

work, wooden spoons made for table use and almost as fine as horn. I have a long slender punch dipper made in this district; the slender handle, 9 inches long, is barely a quarter of an inch in circumference. There is a small carved notch near the end to prevent the spoon from sliding into the bowl.

It is as finely finished a piece of work as you would wish to see.

Heavy wooden dippers to hang on to the side of the pig swill barrel are naturally coarser, but quite as well balanced.

Most of the wooden spoons made in England are the work of bodgers and very characteristic of their localities. In the North you get the straight-handled spurtle. A *spurtle* is so individual that it deserves precedence. In its largest size, it appears in the old ballad:

> What's this now, guid wife?
> What's this I see?
> O how cam' this sword here,
> Without the leave o' me?

> ' A sword,' quo she;
> ' Ay, a sword,' quo he,
> ' Shame fa' your cuckold face,
> And ill mat ye see;
> It's but a parritch spurtle
> My Minnie sent to me.'

> ' Weel—far hae I ridden,
> And meikle hae I seen,
> But siller-handled spurtles
> Saw I never nane!'

A spurtle is straight, slender, and very sturdy; it has no scoop at the end like a spoon, but a straight bottom edge that will scrape well across the bottom of the pot to prevent the oatmeal sticking. It is interesting to note that the most modern spoon on sale in the newest scientific cookery departments is nothing but an old-fashioned wooden spurtle, used through the generations for stirring porridge and jam and stews.

Where peat is burnt (both here and elsewhere) I have found the indigenous wooden spoon designed with pointed-handle or ' tail', because it will jab upright into a peat and stay put.

What to do with a spoon when you take it out of the pot or dish is always a tiny problem: it's sticky—you can't lay

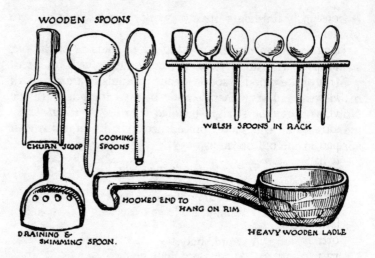

WOODEN SPOONS

CHURN SCOOP

COOKING SPOONS

WELSH SPOONS IN RACK

DRAINING & SKIMMING SPOON.

HOOKED END TO HANG ON RIM

HEAVY WOODEN LADLE

it down. Therefore, in Wales, they make small spoons, with pointed tails, which hang up in rows in a rack, and hooked tail spoons which hook over the side of the barrel,

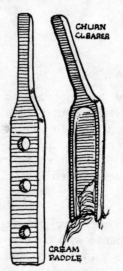

CHURN CLEARER

CREAM PADDLE

or churn. As a development, this hook end is often made square and as, for constructional reasons, it must be originally the same depth of wood as the bowl piece, it is often left this full depth, so that when the spoon is laid down on the table, the handle and hook form a little strut and the large spoon becomes temporarily a small steady bowl.

A set of pointed-tail cooking spoons, hanging tail down, in a neat rack with her name on it, was a very significant gift from a country sweetheart; it expressed his intentions nicely.

The characteristic wooden spoon of Devon is now very rare; it had a thin flat edge for sliding under the cream. I do not know if these are still made in England: they were as beautiful as tree leaves.

In the workshop near Llandilo were also the remains from

22

a consignment of butter hands, and several butter stamps. Butter sold in commercial quantities is usually moulded in rectangular shapes for convenience in packing, but where small quantities of butter are privately sold from local farms, the round pound is preferred. These moulds consist of the bowl and the stamp. The butter is pressed lightly into the bowl, the stamp, which should fit the bowl exactly, is laid on top and given a sharp rap, then the butter is turned out bowl shaped, with a handsome pattern on top. The patterns, now used haphazard, used to be very specialized and important. They were the hall-mark of the farm and usually characteristically carved. A valley farm might show a ' Swan and Rushes '; a hill farm a spray of ' Bog myrtle ' or ' Gale '. A farm with a tradition for pedigree-breeding might have a portrait of a prize-winning cow, repeated in golden butter every market day.

§ 8.   *Coppice Work*

Coppice work deviates from the heavier ' treen ' in that the ' camps ' are usually lighter, as the coppice workers move more frequently, and coppice products are usually fetched away without delay. The shed for bodgering jobs may be left standing the whole year, and be used for a ' seasoning ' shelter in which to store the goods after the workers have finished, since the lorry only comes once, at the completion of the contract. But coppice work is quickly ' cleared ' and the lorry comes and goes nearly every day.

Coppice is young quick-growing wood, planted close enough together to grow straight, and cut at various stages of its growth and at special times of the year. Sometimes coppice is cut to make palings, sometimes for poles for young fruit trees or rose standards, or for gates or fences. Usually the wood is graded as cut, so that it can be used for two or three purposes, according to its size (Illus. 6).

The method varies, but usually one worker goes ahead, spacing out and clearing. With an axe he fells the saplings into piles, and then the next workers follow, sorting, cutting, splitting, and ' wiring ' each pile, that is, binding into bundles or ' packs ' for transport. If the work is to be continued in the coppice, the wood is carried to the shed ; otherwise, when it is stacked and ready for the lorry, the coppice work is over. The last worker is the solitary man who, having prudently annexed a few spare stakes and purloined a length of wire, mends the fences where the lorries have broken them. Then he walks ponderingly over the ground, looking about for what has been lost, putting the old camp kettles into the hedge. . . . Then he, too, drifts away, and the empty coppice is left to the squirrels, the birds, and the bare black fireplaces, over which the little moss comes creeping back. . . .

You can make coppice of oak, ash, hazel, chestnut, or ' mixed coppice '. Large established concerns, such as hop fields or fruit orchards, often own and grow their own coppice and it is cut by their own men. Otherwise coppice is contracted for and ' bought standing '.

Larch and fir are not coppice ; they are plantations.

Tough pine logs are staple usage for pit props in mines. Therefore, before so much mine timber was imported, most old quarries, rough land, and hill estates were planted for props. The timber was sold at standard growth of 15, 20, 25, or 50 years. But now the price of the wood is too low to be able to pay for re-planting.

The huge government afforestation departments now cover this country industry, and not only are the wide water-work and reservoir areas planted with fir, but many wild moorlands and mountain slopes are being fir forested. In this ' plantation ' industry, new long nursery slopes are diapered out with tiny seedling trees. Drains are cut in the peaty soil, and fire-breaks stripe the hills.

It may be of interest here to describe the work I saw in progress in a hazel coppice in Kent, where they were making split palings. It grew on the crest of a chalky hill-side, very straight and clean, and in fine condition. The weather was very cold : March winds whistled down the valleys and sang between the hop wires, almost cracking the flints in the fields. But cold weather is an advantage for cutting hazel, for a small growth below the bark (Hazel weevil or *Curatio*

24

coryli) may occur in May, or even earlier in mild unseasonable weather, and the wood is then useless.

Up on the hill-side the bleached primroses trembled stiffly, and there was crackling cat-ice in the cart ruts. The freezing weather had dried every scrap of moisture off the wrinkled surface of the earth. Starving birds followed me, for my footsteps broke the ice on the tiny frozen woodland pools and they could get water. One deeper puddle, where an inch of water and mud showed, was ringed with small birds within a few minutes. The willow catkins looked very hard and tight, and the curled crisply-thin beech leaves rustled and scratched in the bare hedges. In a sad little grey bundle under the leaves was a frozen squirrel; perhaps his winter hiding place had been broken by the wood-cutters, or perhaps he was too old and weak to live until another spring. Farther on was a stoat, quite still as he had never been in life—his small strong legs stretched out, as if he had gone on running and leaping into the next world. The whole landscape was frost bleached, colourless, and flat, even the black tarred Kentish barns had a dull bloom over them, like the colour on a sloe in November.

As I climbed the bleak hill-side, the wide square Kentish fields spread out below me : the ' root-fields ' were being eaten off by huge-boned Kentish sheep, who seemed to move slow as elephants over the grey earth. In one field was a heavy Kentish plough, with its astonishing six-horse trace, spread out like the sea anchor of a Noah's ark, waiting for the thaw.

I had seen a lorry swing along the Maidstone road with new and very white wood on it, so I knew they were ' cutting ' somewhere around, and struck up inland to find the place. Suddenly I met the scent of wood-smoke, sharp in the frosty air. Half a mile up wind, I found the still hot ashes of the woodman's fire. I sat down and warmed myself, and then walked around and looked at things, and then leaned my back against a tree and made the diagram on page 27.

It was a three days' clearing. Three fires lay spaced out in regular line up the hill, each fire a perfect circle. The earliest was black and empty now, and over its rim the growing things were feeling their way with tiny green tentative fingers.

25

The second fire had black charcoal lumps left in its centre, and a white ash rim nearly all blown away. The third fire was still hot white fluttering ash, and over it, the heat haze hung quiveringly, so that looking across it the trees beyond swam mistily like a forest under water. Each fire had burnt out into a perfect circle beyond which the charred twigs radiated, smoothly, like the spokes of a spinning wheel.

It was dinner hour, and no one about; a bicycle leaned against the hedge. No houses were near. In the lorry tracks, the short grasses were springing up again; so I knew that the lorry had only just gone and had probably given the workers a lift home to dinner, and would soon be bringing them back. I saw, too, that the first clearer had left his felling axe; it lay sideways near the fire, uncovered—he would have covered it against the frost if he had been leaving for the day. His hand axe was under the leaves at the foot of the tree where he had been working. Only one man was clearing; but two were trimming.

I was still drawing when the workers returned. The lorry dropped them above the wood, going on to a further camp, and would stop on its return.

The first woodman started to get on with his clearing, and talked as he worked. He had been abroad, to the West Indies and China and Canada. We travelled in talk, via Fre'mont and the Rockies, back to the Kentish coppice, speaking of the grey squirrel. Like most wood-workers he had a poor opinion of this invader. The spinney was rapidly coming down and sorting itself out into orderly piles as he talked. He flung the trimmed boughs aside into two piles, thick and thin, on either side, and the trimmings piled up, ready for burning.

Lower down, the splitting was going on. There is a knack in splitting. They split with a knife fixed into a stump. The next worker 'length'd' the split poles, cutting them down to size and pointing the ends with two blows of his axe, dropping them, as he did so, into a barrel sunk in the earth. This allowed the bundler to grade them accurately by withdrawing the poles selectively, any slightly longer or shorter ones going into the same bundle. Willow rods, ladder rungs, dozens of small woodland jobs are simply 'barrel graded' thus; the

26

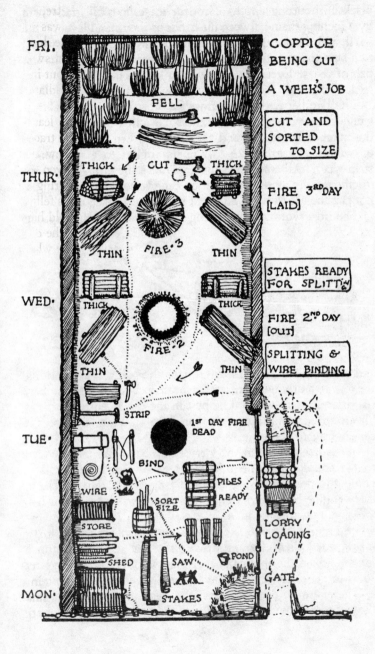

FRI. COPPICE
      BEING CUT

      A WEEK'S JOB

      CUT AND
      SORTED
      TO SIZE

FELL

THUR· CUT THICK
THICK

      FIRE 3RD DAY
      [LAID]

THIN FIRE·3 THIN

      STAKES READY
      FOR SPLITTING

WED· THICK THICK

      FIRE 2ND DAY
      [OUT]

FIRE·2

      SPLITTING &
      WIRE BINDING

THIN THIN

STRIP

      1ST DAY FIRE
      DEAD

TUE·

BIND

      PILES
      READY

WIRE SORT
      SIZE

STORE

      LORRY
      LOADING

SHED SAW POND

      GATE

MON· STAKES

blacksmith cutting lengths of iron for horseshoes will grade them by bunching them between his hands as they stand level on the anvil, and the doorman picks them out in pairs—two longest, two next longest, till the two smallest are left for the smallest pair of shoes. Every country worker knows this 'pairing out' trick.

The bundler had made himself a most ingenious packing-gauge from two zinc pail handles. He piled the stakes in till the notch was reached and then wired down with the lever, or woodman's grip vice (which is similar to a mechanic's strip-vice). All woodmen use tools like these, making them for themselves as required; they cannot be bought in shops, nor can the use of them be learnt in books (Illus. 7).

The post workers were using the heavier wood, stripping

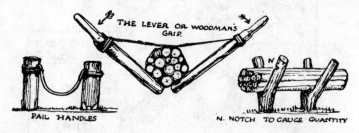

THE LEVER OR WOODMAN'S GRIP.

PAIL HANDLES

N. NOTCH TO GAUGE QUANTITY

and squaring the posts to length (for the palings) and cutting the foot to a driving point with a few axe cuts. In most timber, the driving point will not be best in the exact centre as that is 'heart wood' and more soft than fibre wood, so the point is set slightly to one side.

In the late afternoon the lorry came back and manœuvred to the coppice side, negotiating two cows and a ditch, and coming to position just against the hedge. As camp kettle and lorry engine boiled simultaneously, there was a break for tea.

Then the procession of workers started, taking the new white bundles of palings down to the lorry, where the lorry man loaded them up and drove off. He took with him a message from the foreman ' that they would be working later so as to get clear for Saturday '; and gave me a lift home.

There are 11,627 foresters and woodmen given in the Government Blue Book, and 172 of them are getting on for 80.

§ 9.   *Stack Centres*

He grippet Nelly hard and fast ;
Loud skerl'd a' the lasses ;
But her tap-pickle maist was lost,
When kiuttlin in the fause-house.

Burns

There are many names for these wooden frames, 'boss', 'stack-centre' (lowland colloquial Scotch), 'false house,' or 'truss'. They are either made on the farms where they are used, or by the local woodman.

In England where they are only used in wet harvests or where a circular stack is being set up in a meadow, they are

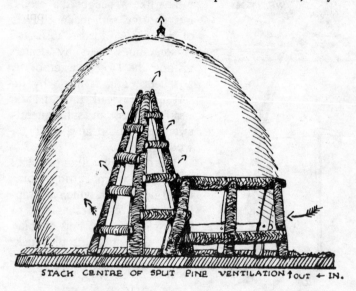

STACK CENTRE OF SPLIT PINE   VENTILATION ↑OUT ← IN.

most often knocked together on the spot, of rough timber. But in the North and West where the climate makes them of more general usage, they are a very distinct country 'article of trade'. In the summer, in the Carse of Gowrie, you may see hundreds.

The idea of the boss is to ventilate the stack. The natural heat of the drying hay rising, causes an indraught that sucks up the cool air from the ground level. All stacks breathe thus in drying, and these centres form an artificial respiration chamber that facilitates the natural process. They run about 6 feet high, varying with the size of the stack. Sometimes a large stack has a little low shaft leading into its centre.

§ 10. *Walking-sticks*

The walking-stick must have a word to itself. Ash is the 'literary' favourite, but thorn is used quite as much by the country people when cutting a stick for themselves. The 'crummock of the Isles' is rather like a shepherd's crook, but shorter and made entirely of wood. A 'loaded' ash staff is sometimes used by dealers visiting the Hog Market or by poachers. This is made by inserting soft lead into the slit head of the ash, and subsequently swelling and pegging over the wood.

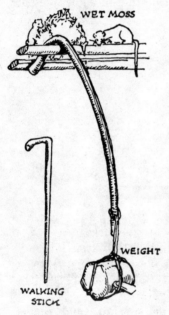

WET MOSS

WEIGHT

WALKING
STICK

The ordinary walking-stick of ash is cut when cutting down thorn hedges (in which ash saplings are great trespassers) or when clearing a young coppice. When they are cut, the sticks may be somewhat bent and twisted, but while green they are hung up from a rafter in the barn, with heavy weights, often lumps of rock, attached. Sometimes they are steamed over a fire, but care must be used to prevent the outer bark splitting, for this will spoil the appearance of the stick. A rubbing on the grindstone finishes the cut at the top and smooths off the marks of any lopped side-shoots. The ferrule is usually supplied by the retail firm to whom the sticks are sold.

Country folk say about the opportune moment: 'I does

30

it what time the old man cuts his walking-stick,' for once upon
a time; a young man asked, ' Grandfather, when do *you* reckon
is best time to cut an ash stick ? ' and the old man replied,
' When I sees it.'

§ 11. *Hop and Fruit Poles : Tar Dressing*

In hop and orchard country the young trees need staking,
and in these districts the preparation of the stakes keeps the
workers busy all winter.

Stakes are usually cut from local coppice ; young chestnut
(sometimes hazel) is lopped, cleared, and sorted into lengths,
and by February the ends are tarred and the young trees staked
against the winds of March.

I came upon one particular tar tank in a Kent orchard. It
was in the centre of a valley, where the sloping hillsides
spread outwards and upwards on either hand, like a wide open
book. On the right slope the winter sunlight slipped fitfully
in and out of bare orchards, printing long thin shadows on
the ground. On the left slope stood serried lines of winter-
bleached hop poles, gaunt against the white flinty earth.
A great team of huge black horses was clanking homewards
beyond the white poles, their steel traces shining white as
silver.

The tar tank was brick built. It held a small furnace 12 feet
long and a chimney 14 feet high. A dump of coal, and a fire-
screen of a sack and two sticks marked the fire-place. The tar
was water-covered against fire and was still warm, and on its
iridescent surface (coloured like the breast of a black pheasant)
was floating a log, whose weight together with a ' gate '
of staves held upright the soaking poles.

The poles may be anything from 5 to 20 feet in length,
but that day they were chiefly 12 feet, because they were ' cut
downs ' : where hop fields adjoin fruit orchards, the old tall
hop poles are cleared, and then cut down into short tree stakes.
The long hop poles are chiefly of Norwegian larch. They will
stand thirty years and more if properly treated, and when
removed from the hops can still be cut into three or four
lengths, long enough for support for the young trees.

31

This was the job ' Tar man ' was doing that day.

The poles must be left to get cold *in* the tar, the longer the better. The poles and stakes must be set upright because, in tar-dressed timber, the wood must be immersed *from the ends*. As it heats up, the sap is driven upwards and bubbles out at the top, and the wood should be left to cool in the tar so that the heat expanded cells suck up their fill of tar. If the timber is dropped into the tar, so that both ends are immersed, the air inside the wood cells is imprisoned and the preservative cannot penetrate.

Sometimes the tar tank is used for dressing the felt and cloth pads that are put between the tie and the post to prevent bruising the bark of the young trees. The sap-stream flows close beneath the bark, therefore, if you cut through the bark in a ring, complete around a tree, it will die. In some cases this is done before felling on purpose to kill the tree and drain it dry. Rabbits eat off the sappy bark in a circle, so that a pest of rabbits can destroy a young orchard : hence the wire netting around the base of fruit trees. The straw rope used for ties is described elsewhere (page 76), and pads and ties are also made to-day of such irrelevant things as old motor tyres.

Tar dipping is not an important process, but the little brick tar-tanks, usually made by country people for country people's use, are so simple and efficient that they deserve this passing comment.

§ 12.   *Forestry Camps*

Between the carpenter and his workshop and the camp workers of the woods come the large fixed forestry camps, settled in one district for many years, using both local and imported timber. These settlements partake of the nature of both factory and camp, since there must be a fair amount of building to house the heavy machinery, and yet much of the work is carried on out of doors.

There is a good example of such a settlement on the east slopes of the Forest of Dean. They have a heavy rack saw for dealing with large timber. The rack is the iron platform which travels forward, carrying the log ast the circular saw. (These

larger circular saws are now fitted with replaceable teeth—
a Canadian device developed since the War.) The workshops
also have heavy lathes and drills, an engine house with drive
and carrier belting, and other plant, as in a small factory.

In large camps, once the initial output for plant is worked
off, the running expenses are comparatively light, but whereas
these workshops once used to employ a fair number of regular
workers, and had boy apprentices (who, doing small jobs, had
time and opportunity to learn the work), this particular camp,
like many others, had now reduced the staff to a bare handful
of trained men, with no apprentices, because though the work
was not actually less it was too irregular to employ many
workers permanently.

We cannot compete in growing timber with the great forest
estates of the Northern countries. This type of firm, therefore,
is put into the curious position of being either out-of-work
or over-worked. Manufacturers, who often make their big
profits from large consignments, manufactured abroad, now
use these small English firms for their convenience on ' odd
jobs ', emergency orders, special sizes, samples, or for completing consignments when there is need for haste. Any
unfairness lies *not* in foreign competition but in conditions of
sale. This commercial buyer, thinking only of profit margins,
expects to pay the same proportionate price for the small
English order, rushed through at his own convenience, as he
will pay for the large order, bespoken beforehand and executed
at leisure.

Actually, this spasmodic type of employment, thus forced
upon the smaller country workshops, involves them in serious
difficulties. Lacking the huge capital of the larger concerns,
they cannot afford to keep a large staff waiting between jobs ;
so there come to be fewer English places able to cope with
orders, fewer skilled men employed and (most serious of all to
the countryman wise in nature's plan), ' nothing to spare for
apprentices ; no lads learning.' There are ' no young ones
coming on, *able* to work when we old ones go ', said another
workman shrewdly, ' by the time all these trees that the
Government's planting be ready for felling, there'll not be
nobody left to work them.'

33

Formerly the worker chose his raw material and fixed his own selling price, by the known goodness and worth of the goods he had made: the maker was directly responsible to the customer for the value and endurance of his wood. Nowadays, the maker is far removed from his ultimate customer, and the ' Business Heads ' of wholesale and retail firms are far removed from the workshop and timber yard. They demand ' appearance ' and cheapness; as to fantastic timber yard pleas for durability, the sooner the goods wear out the better ! For then the customer must buy again.

Assuredly, it is not only the value of the goods, but the integrity of the English workmanship itself that is preserved in the country workshop. Yet, sadly like the sound of steady felling axe through the woods comes again the echo of the old woodman's words, ' Maybe, by the time all they trees they been planting be ready for felling, there'll not be nobody left to work them.'

§ 13. *Carpenters and Joiners*

The *Carpenter* works with the architect, all the fixed and essential parts of the building are his. The *Joiner* comes in where the ' fixtures ' and furniture begin. The *Cabinet-maker* does fine smaller work. The work of the typical country carpenter defies description in general terms, for he may be called upon to make anything from a china cupboard to a chicken coop.

Take, for instance, the work of a small country workshop in my own experience during an average week. Some of the jobs had already been begun and some overlapped into the week after, but they may be said to be fairly representative :

(1) A heavy oak garden door, about 6 ft. by 4 ft., nail studded, and re-slung on the original old iron hinges; also the jamb and lock case. This had been made to the specification of an architect, and a very good piece of work it looked.

(2) A wheelbarrow new bottomed, dried out, and repainted.

(3) A well head and cover made for the local council. (This carpenter used the roller from an old mangle for the

winder.) He had completed the wood-work, but was waiting for a second man to help him to fit and cement the head into the brickwork.

(4) A beautifully made, neatly dove-tailed box, with measured and shaped compartments, lined with baize, for the church silver, complete with carrying strap, and lock. This was done at less than cost price, because he was a church-warden.

(5) A small repair to one of his bee-hives (and probably other small jobs carried forward that I failed to check); also the job I took him myself when this list was made:

(6) Mending an old chair by 'surgical' treatment for dry rot and inserting a new rail. He left me to stain the new rail because he was 'pressed for time to go back and finish the well head'.

(7) Re-runging a broken ladder.

So much for the general work of a country carpenter. Many specialize in fine furniture; others retail some special line, such as beehives, or hen-pens, or carved woodwork. Most country car-penters must do interior *and* exterior decorating, painting, plumbing, brick-laying, concrete work, and plastering.

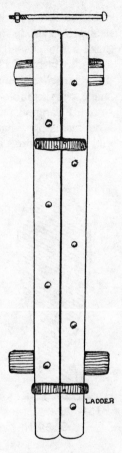

LADDER

SPACING THE AUGER HOLES

It is difficult to select one represent-ative object made by all country carpenters because their

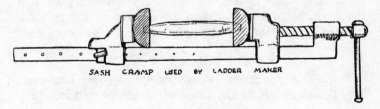

SASH CRAMP USED BY LADDER MAKER

35

work varies with the requirements of the district. But practically everybody needs a coffin, so all country carpenters make coffins, usually of elm or oak.

Coffin fashions change more than one would expect. The older carpenters remember making many that were of fish-tail shape, with a double curve. Now they are usually cut rounded or with 'angled shoulders', or sometimes 'straight-sided'. The fastenings, handles, and linings are bought wholesale and kept in store. Metal plates are sent away to be engraved, or the sign-painter may be called in.

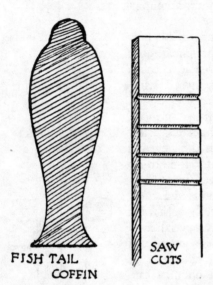

FISH TAIL
COFFIN

SAW
CUTS

A plain elm wood coffin represents a good day's work, even starting early. Where the sides are to be bent, shallow saw cuts are made (called saw calfing) and boiling water or a hot iron used, bending and securing the sides to the shaped base and top. Sometimes the wood is moistened and a little fire of shavings and chips lighted inside the coffin till the curve is 'set' (as a shipwright bends his timber). The planing and sand-papering is done with the greatest care, and if any special 'finish' is asked for, the worker will sit up all night to get it done to the very best of his skill, for the work put into this job is never grudged. It is a more intimate job in the country than in the town, because town coffins are ordered to size; in the country they are made to measure.

§ 14. *Trugs, Slops or Skips*

Slops, skips, or trugs are made of split wood. The wood may be oak or ash, or any wood which can be split into long strap-shaped lengths. The frames of skip, slop, or trug are usually of bent birch, though willow or other pliant rods may be used.

They are perhaps among the most interesting of the things that are made in England. As far as it is possible to judge from historic manuscript drawings this craft is centuries old. On a large scale, it is used for hurdles and makes light wooden walls. Wattle and daub, as its name implies, necessitates wattle, that is, wand or willow open basketry. The daub, once thoroughly well worked (while wet) into the interstices of rough wattle work, was immovable, and walls so made, secured between uprights of oak, could be worked to a fine, smooth finish without any fear of the plaster falling from its grip.

At present, the general use of split wood is for farm baskets and its more specialized use is chiefly for garden paling, hurdling, the walls of garden shelters, and barn partitions. The silver oak weathers very pleasantly in contrast to the darker stakes, and the material is very durable. In modern use, such split wood hurdling is often treated with some form of creosote preservative. The older trug baskets or skips or slops vary slightly in different parts of the country : Sussex ' trugs ' are very distinctive.

I will describe the slops made in a small place near Kendal. The workshop was in an old shed adjoining an old tan-yard. Originally the oak bark stripped from making the trugs had been used in the tan-yard, and probably the leather-like strippings in the tan-yard had been used to finish the trugs. The large boilers and soaking vats in the yard were full of water which was the colour of strong tea and the piles of unused bark accumulated unwanted in the yard.

The frames of birch were made first and hung in clusters of standard sizes from pegs driven into the stone wall. The split oak strips were trimmed and finished in a long ancient-looking foot-vice. This particular vice was of solid oak and must have been a couple of hundred years old. The seat was padded with a folded coat, the worker sitting straddle-legged over this. The straps of oak are wound over and across the birch frame by strong movements of the hand and forearm. The illustration shows the process of weaving more clearly than it is possible to describe. The cross straps go from side to side

SHAVING STRIPS WITH DRAW KNIFE.

FOOT-VICE IN A
NORTHUMBERLAND WORKSHED

of the slop and are held together by the strips which are interwoven through them from end to end. Nothing secures the strips but this intricate and strong interweaving. The ends

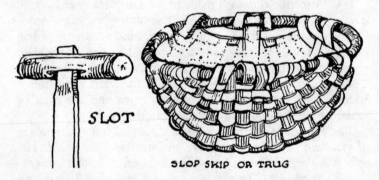

SLOT

SLOP SKIP OR TRUG

are brought up and over, and inserted back under the cross-weave; the worker carries the sizes, spaces, and 'trend' of the straps so clearly in his head, that his hands are really following out a pattern he already sees plainly.

38

There are many variations in shape. There is a right-hand slop or a left-hand, according as to whether the basket is intended to be carried against the right hip or the left. The left is more popular, because a man uses his right hand either for feeding from the basket, or gathering into it. But left and right hand are both needed for market garden work, between rows. A kidney slop is shaped like a kidney, for carrying in front of you, curving round the body. Sometimes they are fitted with strap and handle. Their chief use is to carry fodder about a steading or out to the ewes; a few are used by stone pickers and other

ADAPTED FOR
SOWING BASKET

workers requiring a very strong basket. A large oval trug, or slop, makes a good crib for a baby. It is easy to carry, and rocks a little with the foot; berry-gatherers and pickers often use one and the child they are suckling will sleep in it peacefully beside them as they work.

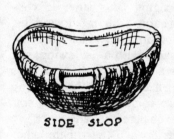

SIDE SLOP

§ 15. *Shepherd Crooks*

These are often of great antiquity, being used in the same family for generations, like the sheep bells which are often the treasured property of one particular family.

A smithy I know in Sussex specializes in crooks for the South Down sheep. Such a crook is about 6 feet long, of ash pole and iron, and wrought more cunningly than is apparent. The slot must just be the width of the sheep's stout little hind

39

leg-bone, and the end curved outwards, and blunted, so as not to jab, and the whole is set at a slight angle downwards. And the crooks made for Southern sheep would not hook their Northern brothers.

§ 16. *Billes*

Make ready now each needful instrument,
. . . also the curved knives light,
In plant young, a branch away to take
The hooks, that the fern away shall bite
The billes all the briers *up* to smite.   *Palladius*

The bille, in general, is made for forearm chopping, as distinct from the swing-axe movement; the heavier billes have sometimes a sweeping sickle cut, but the bille is definitely weighted and designed for the short cut. They are used in coppice work and for all sorts of hedging (this latter in connection with the long bille, or trimmer, which cuts upwards with a half swing, half jerk action). Each man to his own bille; and the man who has found one fitted to his weight and liking is ill persuaded to change it for any other.

STONE NOW HELD IN PIECE OF INNER TUBE MOTOR TYRE.

Only a few of the many types are illustrated here, since more than forty different types are made by firms specializing in agricultural implements.

[A] An exceptionally thick bille, narrowed to keep down the weight, and subsequently sharply set to its cutting edge, which is narrow. It is a northern type. I found this specimen cutting hard ling heath for making besom on the Yorkshire moors above Appeltreewick ; the type is more common in the North.

[B] An exceptionally wide bille, therefore thin to keep down the weight ; the wide cutting edge slopes gently on one side and more abruptly, to stand up to a harder cut, on the other. This bille was in use on the Wymes Wold road, between Leicester and Nottingham, where the wide grazing on either side of the highway has fairly soft grass for the sickle edge, with occasional tufts of thorn and weed for the straight edge. It is a typical Midland type.

[c] Was in use near Chipping Norton, Oxfordshire; the flowery hedge sides there have a strong keck and bramble growth, which accounted for the sudden strong hooking cut at the end.

[D] Is a Kentish bille, a curious compromise between the Northern strong cut and the Midland sweep. It was in use for

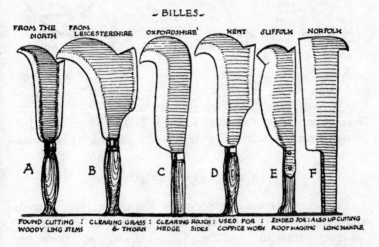

~ BILLES ~

FROM THE NORTH   FROM LEICESTERSHIRE   OXFORDSHIRE   KENT   SUFFOLK   NORFOLK

A   B   C   D   E   F

FOUND CUTTING : CLEARING GRASS : CLEARING ROUGH : USED FOR : ENDED FOR : ALSO UP CUTTING
WOODY LING STEMS   & THORN   HEDGE SIDES   COPPICE WORK   ROOT HACKING   LONG HANDLE

coppice work. It had something of the thickness and weight of the Northern bille, but the sweeping edge was for twigs, not grass, and the cutting edge would take through several inches of wood.

[E] This was sketched in Suffolk, near Pinmill, and seemed well adapted for the rough wood mixed with sedge and rush of the estuary. The short straight bite at the end was for downward, woody root hacking. In contrast to the other billes, the weight was heavier at the end. It was being used on banks and edges somewhat below the level of the worker. The small bend in the haft gave a good lift recovery to an upward stroke and the cutting edge was continued sharply, close to the handle.

41

[F] Was fixed to a long handle, like an up-cutting bille but used with a downward movement also, where clearing reeds that encroached against a hedged bank by a Norfolk roadside.

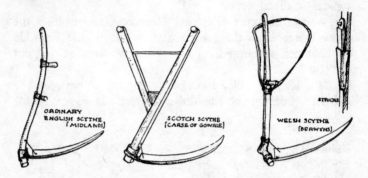

ORDINARY
ENGLISH SCYTHE
(MIDLANDS)

SCOTCH SCYTHE
[CARSE OF GOWRIE]

WELSH SCYTHE
(BERWYNS)

STRICKLE

A butcher's chopper is not called a bille, but a cleaver, and again varies in type, according to the work.

All tools made in England vary according to locality. Scythes have almost as many forms as billes. The illustration shows three examples, all in use to-day in various localities.

§ 17. Coopers' Work

This work is most interesting and has its own special tools and traditions. The skill is of hand and eye in binding and wedging, rather than cutting. Both iron and wood hoops are

BUSHEL & SKIRE.

used. The wood hoops and sections are often supplied to the coopers by the coppice or wood workers, but the coopers carry on their trade nowadays chiefly in town workshops so a description of their craft is not given here.

§ 18. *Clogs*

Clogs are now ceasing to be a mass-production article and are becoming again the work of the smaller country maker. The closing of many pits, and the changes in employment in the mills have cut down the enormous output of clogs that used to clatter to and from work in the North, but there is a definite demand for a certain number of clogs for special types of work, and they tend now to be more specialized for the workers' particular needs. Among the present users of clogs are pit workers, mill workers, dairy workers, farm-hands, and any mechanics whose work necessitates their

SCOTCH HANDY

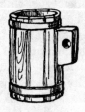

OAK MEASURE

standing on slats or draining boards before their machines. Even the thickest leather, when continually damp, becomes bent and uncomfortable, while the wood stays smooth. Yard workers—on cobbles—whether on hill farms or in the dray horse sheds of a railroad or quay, use clogs. It is characteristically a 'stone' shoe, not a 'sand' shoe. In towns, the culvert, drain, and underground workers use a thick rubber and wool-lined clog, adapted for their special requirements.

The Yorkshire, Westmorland, and Lancashire clogs all vary, slightly, but definitely. Leicestershire and South country clogs vary from those made in Cheshire, but there are comparatively few clogs made or worn in the South of England.

When required they are usually got from the North through some manufacturer who specializes in the industry for which the clogs are needed.

Neither oak nor ash, nor any wood that splits can be used for their making. The gypsies used to monopolize the clog sole trade, cutting the blocks into rough shapes and selling to the finisher. Sometimes the gypsies sold direct, sometimes the clog-maker bought the trees and commissioned the gypsy to camp there and do the cutting. Under this contract work you will see a scrap of the bark left on every block, to show that the gypsy had not wasted wood by cutting the block smaller than needful.

Clogs are born as twin soles, and live in pairs from the moment the block cutter shapes them. Thus you can be sure the shrinkage will be exactly equal, as the grain falls symmetrically either side of the cut. He uses odd strips of leather or the outer bark from stripping osiers, and two small nails, to tether the two soles together. Clog soles are cut with a very subtle curve, with a small rim round them to take the leather. The iron sole of the clog should really come under the heading of metal work, but is more conveniently considered here. Large quantities of ready-made clog irons are turned out from town works, but with the present smaller output and better quality of clog, the old made-to-fit iron is coming back into its own. The making of these clog irons is a country industry, often carried out in connection with a local forge or smithy, when other work is slack. I saw it in progress in a small works near Silsden in Yorkshire.

The illustration of the clog iron best explains its subtlety. The lengths for sole and instep are cut separately from the heel-piece. The heel-piece is a comparatively simple affair on horseshoe principles, but the instep piece is very carefully designed. There is a certain amount of spring in the wood of a clog, but no bend, therefore in walking, the curve of the step, as the foot rocks down and up off the ground, must be given by the iron. The iron must also grip the ground and have a flattened under-surface, on which to stand (' otherwise a chap would be always rocking like ') and the transition from this level stance to a forward movement of the foot must not be sudden. The thickness of this central part is varied some-what according to the type of clog; a man whose job keeps him standing about in water needs it to be a good height to keep the sole of the wood above the wet. The length of it depends upon the length of the clog sole, because a tall man is likely to have a longish foot and he likes a level reach when walking. The shorter clog of a little man has a proportionately shorter 'raise' and gives 'a more rounded rocking' walk to his shorter stride.

Nailing these clog soles to the wood also needs consider-able skill. The town-made commercial article is simply hammered on with nails driven into a shallow groove in the sole, and as soon as the sole wears down to the head of the nail, the head of the nail wears off and the sole peels off. Therefore, in a properly made clog sole, the iron is widened and sunk into a deep protective hollow;

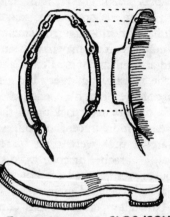

TWO VIEWS OF CLOG IRON SHOWING THE SUNK NAILS & RAISED SECTION

the nail-heads driven to the bottom of this hollow are pro-tected by the iron above. The nails are clamped through, like horseshoe nails—they will never come out till the iron wears flat with the wood, thin as a razor blade, and it will cleave to the sole to the last. The cut ends of the sole are riven up into the wood for extra security, and also to prevent their catching under anything and tearing up. The toe point is flattened and turned up well to give a flat, protective end.

§ 19. *Charcoal*

" Once upon a time, in a tiny hut in a green forest, there lived a lonely charcoal burner who . . . "

Old Fairy Story

The charcoal burner, whom I found within fifty miles of

London, was working in a green wood full of dusk and blue-bells. A nightingale was singing, the smoke from his hearth drifted up through the shadowy trees, the age-old sound of splashing water and the smell of burning filled the air. He told me he had been a charcoal burner all his life, and his father was a charcoal burner all *his* life. When he was a grown man and ' thought he knew a bit about charcoal ', there came an old chap of about eighty, and ' taught him something new '.

The most prosperous days of charcoal burning were probably during the iron smelting of the Weald ; but almost as much charcoal may have been made in the days of chain mail and armour making. Before the use of coal, every horseshoe made in England, every broadsword or coulter (the cutting part of the plough), needed charcoal. It was also much used as a preservative packing in the Middle Ages and later.

At a more recent period, when hops were first grown, charcoal fires were used for the oast houses. Afterwards coal, anthracite, or coke were substituted, but when the oast houses were altered for hot air drying, then a proportion of charcoal was used again.

Nowadays, charcoal has comparatively few uses, but for certain purposes it is indispensable. It is used in aeroplane manufacture, in welding, in making jewellery, in soldering of special sorts, in dentrifice, biscuits, gunpowder, distilling. Artists, butchers, florists use it. Great sackloads are sometimes ordered by Water Boards for use in reservoirs and smaller quantities by country houses for their rain-water cisterns.

The method of making charcoal is one of those interesting compromises between the unscheduled growth of life and the mechanical perfection of a scientific calculation. To seal up and calorify wood in a steel furnace does not produce the same ' quality ' of charcoal. Even if you were to cut the wood till it was all of level size, as even as matches in a matchbox, there would still be some natural incalculable change of weight, of density, in the change of the sap in the wood and the heat of the furnace. With wood, you are calculating in reference to a live thing, not a dead ingredient. Out in the woods the charcoal burner works for mechanical perfection with an instinctive adjustment of this natural variation in values.

The actual procedure of each charcoal burner varies. No two people ever build an ordinary fire quite alike, and this is a very carefully constructed fire. A slow, even heat is needed.

In the illustration, I reproduce the construction of a hearth I watched being built last year in Kent. A pit is sunk in the earth and the logs, brought from the wood and flung down conveniently to hand, are piled up sloping towards a wooden centre which is set open, like a ventilator, in the middle. It is in this building of the hearth that the skill and experience of the charcoal burner is shown. The weight of the logs must lie

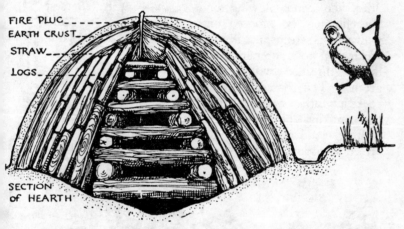

FIRE PLUG
EARTH CRUST
STRAW
LOGS

SECTION of HEARTH

evenly so that the pressure inwards is evenly distributed. This large, heavy log must go there, that thin one there—with an almost subconscious memory of the weights he has already piled in, the worker goes round, layer after layer, till the interior construction of a well-built hearth remains as clear in his mind as its outward layer is visible to his eye. You realize this visual knowledge of the structure he has built, as you watch him going round, moving a log here, to substitute it for a slightly heavier one there, pressing in an extra thickness at one place, and saying (hours afterwards, when the whole structure is buried from sight under smooth earth), 'Put on a bucket of water here, it's here she'll break out, if she does anywhere'. He 'sees' from inside the hearth (Illus. 8).

47

Once the timber is in place, a covering of straw is put on, and over the straw the earth which has been dug out to make the pit. When several hearths have been built on the spot, there accumulates a mixture of charcoal, straw, and baked earth, which forms a little ridge round the hearth. This the charcoal burner sweeps into a smooth surrounding ring. Then he lights the fire down the centre of the stack, closes it over, and watches.

The hearth burns nearly a week from building to finish. At one time they are fierce, later sluggish. As the earth dries, the wind blowing on the pile may be blocked off with rough screens, usually of sacking on poles, or perhaps a hurdle threaded with evergreens. The charcoal burner cocks a weather-wise eye from the smoking hearth to the drifting cloud and moves his screen accordingly. The small

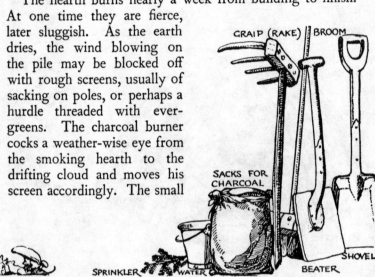

hollow where this particular charcoal was being made was so sheltered that he needed no screens.

Water is kept handy for damping down a hearth, and by judicious management the charcoal burner arranges his work so that he gets time to eat and sleep. But every hearth develops its own temperament, and the charcoal burner must stay within call. That is why charcoal burners usually still live in a little hut by their hearths, tending them day and night.

When the burning is finished, the hearths are opened, the outer crust pulled off with rakes, and the pile allowed to cool. In a well-built hearth, the straw covering will be seen, every straw in place undisturbed but now turned gun-metal

48

black, and as the first gust of wind blows it crumbles to powder.

The great logs and lumps of charcoal shine like starlings' feathers, and crackle and sing as they cool. When cold, the brittle charcoal is packed into sacks, collected by lorries, and taken off for sale.

BARREL MEAT SAFE

Because they live for some weeks in the woods, the men invent all manner of gadgets for their comfort and amusement. A small example, the 'larder' or 'safe' in which they keep their milk or their dinners cool in hot weather, is often found in the colonies ; a box, with one or more sides made of double rabbit netting, sandwiched with charcoal. A basin of water with a hole is balanced on top, to drip. Sometimes barrels with a netted end, as the sketch above, are used. I have seen beer chilled by putting it into a loose bag of straw and setting the boy to swing it in the wind. I wonder how many of these 'woodland' inventions go to the colonies? (*Cf.* drawing on p. 113.)

CHAPTER TWO: STRAW, REED, GRASS, AND WILLOW

§ 1. *Thatching in General*

Theaker, theaker, theake a spanne,
Come off your ladder and hang your man;
When my maister hayth thatched all his strawe,
Hee will then come downe and hange him that sayeth soe.

THATCHING IS so essentially the work of country people, *for* country people, that a discussion of its varied forms is well justified, though thatching is 'done', rather than 'made'.

The ordinary town-dweller is very ignorant of the variety of our thatch. I remember a film producer wanting me to find him 'a haystack for a scene'. When I asked for the locality of the story he looked at me in utter bewilderment, then said, 'What does it matter?—aren't all haystacks alike?'—not knowing that there are at least twenty different types of haystack in England. In fact, there are seldom two haystacks in England made absolutely alike. There are roughly as many different styles as there are counties, and in each county there is the sub-division of material, and the very marked individuality of the maker. About seventeen materials are stacked: hay, oats, wheat, barley, meslin, St. Foin, beans, field peas, haulm, reed (long and short), bracken, occasionally ling, and thrashed straw—and there are probably other stacks (not counting clamps of root crops).

Stacks and ricks are thatched with reed, rush, straw or rye, oat, or wheat, and Wiltshire reed or Somerset reed (which isn't reed at all but specially grown wheat straw).

The stacks are either high, circular, and pointed, or low and humped, or every variety of squared stack. The thatch is secured in different ways too, by rope, spic, peg, bramble, withy, and binder string.

It will be well, therefore, to consider the thatches made in England under various headings. The main division is, of course, between house thatch, which is enduring, and agricultural or field thatch, which is only required to last a season or so. For houses, thatchers prefer straw, rye straw, if possible; if not, unthreshed wheat or oat straw, reed (Norfolk or Kentish) rush, and ling heather. This last is used much in North Britain,

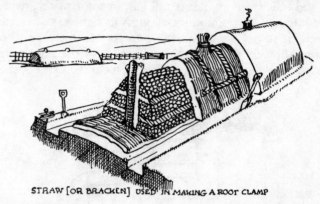

STRAW [OR BRACKEN] USED IN MAKING A ROOT CLAMP

and forms the roof of all the small stone and thatched houses of the North. It becomes very dark outside and under the weather takes on a sheen, somewhat as reed does, but within, if there is no plaster between the rafters, the peat smoke turns it almost black. It is a good thatch, deep and warm.

§ 2. *Thatching for Houses*

The use of house thatch is very old. That it was used as covering for cottages and huts as far back as there were huts at all is common knowledge, but it is less often realized that cathedrals and portions of great medieval castles were frequently thatched. The somewhat bare look along the eaves of many old churches in the country to-day is due to the fact that the top parapet stands up naked against the sky, while the tiled roof behind is almost invisible. Originally the extra thickness of the thatch would have given a much softer, warmer outline.

We have found illustrations in old manuscripts of cathedrals in process of building; from them it was quite evident that across and over the uncompleted walls and unfinished pillars, the builders used to lay a temporary thatch, to prevent deterioration pending completion.

It is a mistake to think that there is anything primitively make-shift about thatch ; it has been superseded by slate and tile and other roofing, but it still has, in some places, its own advantages. One point in its favour is its lightness of weight ; indeed, it is not always possible to replace thatch by stone roofing, because the old beams will not bear the extra weight of slate or tile, nor will the walls withstand the extra outward

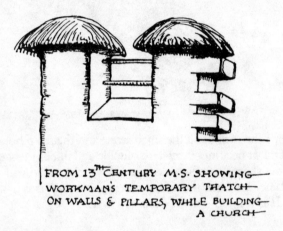

FROM 13ᵀᴴ CENTURY M·S· SHOWING
WORKMAN'S TEMPORARY THATCH
ON WALLS & PILLARS, WHILE BUILDING
A CHURCH

thrust of the pitch. Also thatch is non-conducting. The air under a thatched roof remains at a much more level temperature : the porous nature of the material keeps out both the heat of the sun, and the cold of winter.

In several houses and small cottages the top rooms have had to be abandoned after tiles were rashly substituted for thatch. Where the pent roof is an empty air space, the air fulfils the same purpose of sound and heat insulation, but in the older houses this sloping space was used for the bedrooms, coming directly under the roof. These rooms under a thatch are curiously pleasant and quiet, the thick covering absorbing all noise. The soft sound of rain falling on thick thatch is very different from the patter on tiles or the tintinnabulation on corrugated iron. Spencer wrote the music of it thus :

And more, to lulle him to his slumber soft
A trickling streame, from high rocke tumbling downe,
And ever drizling raine upon the loft
Mixt with a murmuring winde, much like the soune
Of swarming bees, did cast him in a swowne.

The Faerie Queene

One disadvantage of thatch, perhaps the greatest, is its highly inflammable nature. Yet I have known thatched country smithies and blacksmiths' shops which have withstood flames and sparks for years; they were of reed and had, with time, become very closely packed. Large hooks of iron, mounted

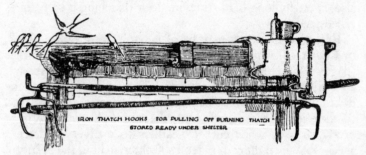

IRON THATCH HOOKS   FOR PULLING OFF BURNING THATCH
STORED READY UNDER SHELTER

on long handles, are hung in conspicuous places in the village street in many thatched districts. Everyone knows of them and they are for pulling off a burning thatch before it falls in and damages the house below. One man can carry the fire-hook to the house, but it takes several heaving on the handle to pull down the thatch.

Some insurance companies either will not quote at all for thatched houses, or make the rate much higher, but thatching material can now be treated to render it reasonably fire-proof.

Near sea or rivers, old fish-nets are often thrown over the thatch, and pegged down. This prevents birds from entering to nest, and is also a protection against gales. These tarred nets are now quite generally used, or fine wire-netting is often put on, and may be tarred also. I have seen thatchers making use of the larger mesh sizes of wire netting *under* the thatch as it distributes the weight from beam to beam, between spaces where smaller rafters have given out, but this is for ' patching and contriving ', not being part of the original structure.

53

Local pattern in thatch is very characteristic; near Hunger-ford, Wiltshire, some cottages seemed very well done, but of unfamiliar pattern; on inquiry I discovered that both reed and workman had been imported from Dorset. Where straw thatch is used the best kind is a round hollow straw, but some threshing machines break and crack the straw. A good thatcher will not use cracked straw if he can get anything else. In Somersetshire ' they do shear their wheat very low and all the wheat straw that they purpose to make thacke of, they do not thresshe it, but cut off the ears and bind it in sheaves and call it reed, and therewith they thatch their houses '. In Wiltshire also, specially long hard straw used for thatching is sometimes called·reed.

Real reed thatching may be of Norfolk reed, Suffolk reed, or Kent reed, while in Wales and the North they use rushes (page 70). Norfolk reeds are extremely long and strong, and will last fifty to eighty years. They are cut alongside the dykes, after the wood has hardened in the stem, and are stacked till the outer leaf comes off. Only the reed itself is used. A light brownish grey when cut, it weathers to a silvery iridescent sheen, like the light on a pigeon's neck, matching the beautiful flint work of the district perfectly, and makes a thatch as smooth as satin. Kent reed is very like Norfolk reed, but Kentish people say that their reeds are harder, seasoning more hollowly in the Kentish climate. In Devon, straw is almost universally used, piled on with the warm thick abundance of Devonshire cream. Oxfordshire thatch is mid-way between the thick thatch of the West and the closer, darker thatch of the Welsh border. Along the Marches we find some of the most individual schools of thatching, small districts each under the jurisdiction of one expert instructor. There is a very neat finish on some of the houses beyond Tewkesbury, which are all obviously the work of one strong and skilled thatcher; and some neat thatch with curious little spikelets, along the ridge in the Forest of Dean district.

On the South Downs, the small villages in the hollows have much good thatch. A large, thatched old wool barn is one of the most beautiful things in England, for the thatch sweeps smoothly up and over the gable-ends in a flowing curve, and then down so low that it almost touches the ground. Some of the under-pinning of those barns are of old ships' timbers, and their curve is the very curve of the chalk. Some districts in Buckinghamshire thatch the tops of the walls. This is also done in Devon to keep the rain out of the cob.

As a rule, thatch divides on either side of a dormer window, which, if it has a high pitch, is sometimes tiled, though the rest of the roof may be thatched. In Leicestershire, and occasionally elsewhere, I have seen dormer windows which were set so far back, rising up through the lower part of the roof, that the window opened out over the sloping thatch.

In the North of England stone and slate roofing are better suited to the climate; but for small mountain huts and buildings further north into Scotland, heather makes excellent thatch, extremely strong and durable. In these parts, as also in the Western Isles, on account of the gales it is frequently roped over, and weighted down with lumps of rock, lashed on to the rope ends under the eaves.

We will now consider some thatching methods in detail.

The illustrations of reed thatching were taken at Lydd, in Kent. The thatcher, a very capable craftsman, was removing most of the thatch completely down to the rafters, some of which had to be replaced, but in another part of the roof he was working over a foundation thatch which was still perfectly good structurally and only needed repair. The reeds were brought to the site by horse lorry in two days' cartage. The reeds had been cut previously, stored, and during slack wet days in winter, ' drawn,' i.e. the soft outer leaves, which had withered and partially fallen, were stripped—in this case by hand— using a wide rake tool. The reeds were then tied up in bundles, each bundle being of a size that could be encompassed by the worker's two arms. They were grounded

DIAGRAM OF REED LOAD

on the barn floor a few times, till they stood level, and the lower band at the foot of the reed was secured first. As they were completed, the bundles of reed were stacked on the wagon, the first layer butt ends out, so that the slope of the tapering reeds was inwards towards the centre of the wagon. This slope was kept towards the centre throughout the loading, so that the subsequent jolting along the road only served to pack the wedge-shaped bundles the more closely on to the cart. They were but lightly tied down (Illus. 14–19).

Under the top layer in the centre were stowed the thatcher's

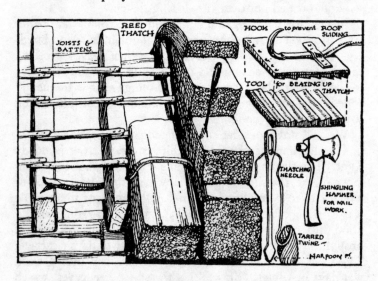

tools. Every thatcher has his own tools, usually made by the local smith to his own requirements, and this one had a short-handled long tooth rake, a Kentish axe, and a beat made (unusually) entirely of iron, with a long hollow socket for the wooden handle. The beaters are more usually of wood, across the face of which grooves are cut at an angle to catch the ends of the reed and drive them back and upwards to a smooth finish. This beater had a hook at the back for hanging conveniently near the worker's hand on the roof. This hook saved him many a step up and down the ladder. There were also a pair of well-sharpened sheep shears, a small bag of nails, rather long, and two thatch needles, one about a foot and a half, the other about 3 feet long. The shortest needle was rather like a harpoon in shape, and thick. He used this needle occasionally for driving downwards through old thatch, the shaping at the point preventing the springing back of the needle in the resilient material. The other needle was a straight, slender rod, plainly pointed. There were one or two bundles of ready-cut split lathing, i.e. half-round small coppice timber for repairs between beams, and a few withies for ties, also a bundle of the doubled withies split and bent into a twisted hook or double prong (see illustration). These, when used in stack work, are called 'spics' or 'speks'. He had a large clasp knife and small whetstone in his pocket, and obtained his ladders and the necessary wood locally. The twine he used was collected from a pub on the way, having been left there ' to order ' by a man who used such twine for mending the fishing nets further down by the coast. This man had sent it up to the pub by another fellow who was bringing barrels the week before. The thatcher was glad to find it when he arrived, as if it hadn't been there it would have delayed him considerably.

It is never wise to strip more of the roof than you are likely to finish in the day, though if the weather is threatening, it is often possible to borrow a rick sheet to peg over before leaving for the night.

Thatching is commenced at the eaves ; in the case of straw a cradle is used to carry the load up the ladder, but reed, being heavier, is usually handed up in single bundles on a fork or carried up by the thatcher, looped on to his back with a piece of rope. The bundles are laid, one above the other, slanting back till the ridge is reached. Each one is secured, both to the rafters of the roof and to the bundles directly above and below it. In the drawing, for diagrammatic simplicity, I have shown these bundles as solid blocks, but the actual process is blurred by the loosening of the bundles and the sweeping upward movement which wedges them under and into the sub-strata of lathing, and ten or twenty bundles would be so used.

Thatch is secured differently by almost every worker, and he must adopt his method to the problem confronting him.

If the roof has been previously prepared for a different type of work, he may have to adapt two or three methods before he can repair it adequately. Where it is possible to strip the entire roof and start afresh, a good thatcher will spend as long with the under-pinning of withy and split wood, as with the surface of reed or straw he fastens to it.

In the roof shown in Illustration 15 the thatch had also to be arranged round the gutter of a dormer window. Shaping of this description calls for skill in design, for the ' lie ' of the reeds must be swept into a steady curve that will bring the ends well down over the eaves and also sweep the length of the curve into the gutter, making as it were a double water-shed.

Where the thatch reaches the ridge-pole, several treatments are possible. With straw, perhaps the most usual method is to bend it over across the ridge by bringing the two top edges of the thatch close together. Reed is stiffer and more brittle and this treatment is not possible, so thatchers usually put a line of reeds lengthwise along the ridge, bring the two top layers of reed well up over this, and sew together from side to side. Subsequently the reed is cut off above the sewing, leaving a stiff ridge, hard and upright, like the hogged mane of a horse. Frequently the ridge pole for reed is plastered, or a light, wooden ridge is fitted over the reed ends.

For this house at Lydd some good thatching straw had been obtained locally, and was used to bend over the length of the ridge ; the yellow straw, in contrast to the browner reed, made a very pleasant if unusual finish. As the previous thatch had suffered considerably from the inroads of birds, this top ridge was at once secured under a length of narrow (rabbit) wire-netting, securely pegged down. The entire roof was later to be covered with tarred fish-net, obtainable locally, but was left to ' settle ' first, as after the first frost it would probably rise a little and need a beating before it could be netted down for the winter.

A small barn, where the thatch was still comparatively good, was re-thatched without disturbing the tarred netting already upon it. On some roofs this process has been repeated so often that, as the thatchers say, the entire thatch, if ever removed, would peel up like a strip of shoe leather, since it is now only secured at the four corners and swung 'same as a hammock, under deck'.

When the thatch is complete, the whole is lightly brushed and patted to a smooth finish. All edges around the eaves and gutters are carefully trimmed; as reed was used here, a beating-in movement was employed instead of the cut-and-trim process needed for straw.

The cottage was of average size; there were two windows, a chimney, and a dormer window. The job took inside a week, including the carting on Monday and Tuesday; Wednesday and Thursday were full working days, Friday saw the work finished and cleared up and included patching the small barn. Saturday, the thatcher went back home.

§ 3. *Varieties of Straw Thatching*

These may be better understood after the reader has studied the next section on the use of straw in thatching stacks. Straw thatching on houses has many special adaptations from the reed usage.

The thatching straw should be as unbroken as possible. Red wheat is a very strong straw, and in some places rye straw is specially grown for the purpose (it is also useful for packing stoneware). The straw is drawn before use (this is described later). If the thatcher buys the straw from a farm or is working at some distance, he frequently 'draws' at the farm in a stack-yard or outhouse, because the farmer can use the refuse straw for bedding, or stack-footing, and it saves cartage. Where he is thatching in the field, he 'draws' in the field, or perhaps in an adjacent pond or stream, and 'rough stacks' the refuse.

I have noticed one very usual variant between the straw-thatchers and the reed-thatchers. Straw is much more dependent on the weather. You will not get a straw thatcher to work in any wind. Also he more frequently has an assistant, probably because straw is lighter and more bulky. It saves a lot of time to have one man below filling the greip with drawn straw and handing up fresh supplies of spic. Two men are needed for putting on lashings, as will be seen in the reference to rick work.

59

Thedrawing p. 61 shows the fan-shaped hold used for holding the straw conveniently on the roof, also a carrier (this one happens to be made from a fork of ash) for taking up the straw. The rake or comb is for streaking down to a smooth finish and the shears for clipping around the edge of the eaves.

The finish of straw thatch, when applied to a roof, is often very neatly designed. Along the ridge where the bent-over straw comes down on either side, this is frequently cut into a pattern not so much for the sake of appearance (your English designer seldom considers ornament without reason), as because the patterned edge is usually dentelled, and more evenly distributes the extra water that falls upon the ridge. If even this small amount dripped down on to the thatching, it would in time wear a line along, across the straws below, because 'rain that runs down, runs off, but if it drips, it strikes through'.

The lines which are carried along the roof from end to end, holding down the straw near the eaves, are sometimes made of very broad-headed spics set in a herring-bone. Sometimes lengths of withy or split wood are laid along criss-crossing in patterns, and held down by spics.

I saw a rather uncommon, but very neat finish put on by a worker near Malmesbury. For lack of withy wood, he had cut some exceptionally long, pliant blackberry lengths, and these he was pegging down over the eaves; this time the material lent itself to a waved line of small curves, rather than the more usual cross-stitch. He told me blackberry was regularly used in his young days, but it was the first time I had seen it myself (Illus. 13).

§ 4. *Thatching a Haystack*

It is now time to describe in detail the thatching of a haystack. The illustrations were taken from an excellent piece of work done in Hampshire, where a haystack was being thatched with straw in a field adjoining the road. There was no pool or stream near, so the zinc cattle-trough used in the winter was filled with water carted from the farm. The straw

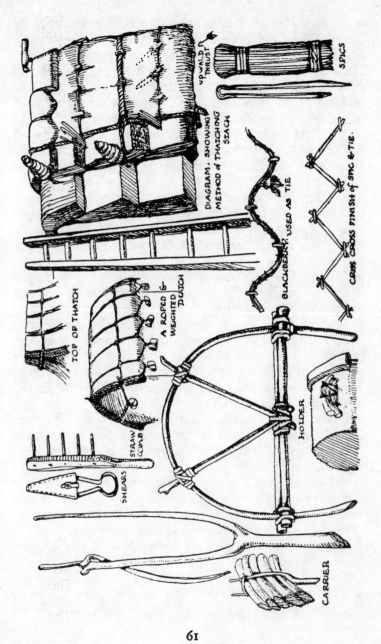

DIAGRAM, SHOWING METHOD OF THATCHING STACK

UPWARD THRUST

SPICS

CROSS CROSS FINISH OF SPIC & TIE.

BLACKBERRY, USED AS TIE

TOP OF THATCH

A ROPED & WEIGHTED THATCH

STRAW COMB

SHEARS

HOLDER

CARRIER

61

to be ' drawn ' was laid under two light hurdles and thoroughly wetted. The thatcher wore the inner tube of two motor tyres down both legs and along one arm. Beginning at the far end of the pile and pacing backwards, he drew out the straw in swift double handfuls with a dragging motion, dropping it on the ground before him, laying as it were a carpet of level wet straws in his track, all lying parallel at right-angles across his path. At the end of the pile he returned, walking forward and picking up the straw, with a level lifting motion, so that as he reached the beginning of the pile again, he had in front of him a large swathe of level smooth straws. This swathe he lifted a little forward out of his way ; then, without turning, he started pacing backwards again, drawing out a fresh supply as he stepped steadily back the length of the pile. This again he evenly collected as he stepped forward, and reaching the beginning of the pile, deposited it on top of the previous gathering. The walk backwards and forwards was the length of the two hurdles holding down the straw, and after he had drawn a sufficient quantity to begin work he had worn a useful smooth track close to the hurdles, and the wetted straw under the hurdles was being swept already into more parallel positions. The drawn straw was then piled into the holder prong, secured by the loop of rope attached to this ; then he lifted it on his shoulder and mounted the ladder to the top of the stack (Illus. 9, 10, 11).

Other materials needed were spics and rope. The spics he had prepared previously, for these are best made in the spring when the withy-wood is young. I have seen spics made of common split pale with a long notch or loose shave cut at one end like a very thin ragged tent peg, but the best spics are made of split withy, and look like a wooden hairpin between 2 feet and 3 feet long, as will be seen from the drawings ; one leg need not be as long as the other.

Quantities of these spics are made odd days in the spring when it is too wet to ' get on to ' the land, and are neatly bundled ' against thatching time '. The bundles are tied at each end, and laid flat along the barn roof or under the corn bin, where they will keep straight till they are wanted.

Nowadays a thatcher will use coir twine for his rope, but straw rope is better for stack-thatching, as it is less likely to cut the thatch (see diagram and note). There is no apparatus needed here at all—the worker simply takes a stick and teases out a handful of straw. He starts his left wrist twisting rapidly, while the right hand works independently, lightly catching and paying out the straws which seem to twist themselves up out of the pile and tighten down under his curved palm to issue —as if by magic—through the loophole left between the thumb and curved first finger as a rope. As the left hand continues its steady motion, the turn of the stick automatically winds the rope into a coil on to the stick. It is a most skilful operation, only to be acquired by long practice. The great art consists in leaving free the loose ends of the straw long enough to entangle and draw out others before they themselves are tightened into the twist. When sufficient coils are made they are laid near the foot of the stack, together with the spare spics and a few stouter sticks for finishing. The folded coat and dinner bag are put under the hedge, the dog settles down on guard, and the thatcher, complete with drawn straw, sticks, and coiled rope, goes aloft to start work. The ' thraves ' or armfuls of drawn straw are lifted and laid on the stack, beginning at the eaves. Here they must project enough to shoot the rain-water clear of the stack-sides. As they are placed, they are spread out smoothly, butt ends down, and at once secured by a bight of the rope, pegged over them from the end of the stack. In this particular stack, binding-string and spics were being used alternatively with straw rope. But the usage depends entirely on the thatcher's own individual style (Illus. 20).

For clarity in the drawing (p. 61) the thatch is shown in diagram-blocks. In use, of course, these become continuous. Each armful is laid one above the other, overlapping about half-way at the lowest point. For convenience, the thatcher's range of work is the width of two ' lanes ' or ' strakes ', that is, the reach of his arm to the right. Thus he covers the part from the eave to the ridge of the stack completely before it is necessary to come down and move his ladder two more ' lanes ' to the left. If an assistant joins him by noon, he will be working sufficiently quickly for the assistant to be kept busy carrying up fresh supplies of drawn straw and spics. It is very important that the spics or hooked stakes holding down the ties should be driven in well *upwards*, as otherwise they would form grooves down which the rain could run and penetrate the stack. If a very steep pitch is wanted (and they are naturally steeper in wet localities) the thatcher lays a 'backbone' of straw lengthways along the ridge, under his thatching.

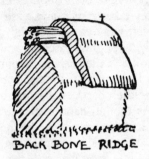
BACK BONE RIDGE

Where cross-over straw ropes are used, they are flung over the top of the stack at the top of each lane, and pegged down by the assistant temporarily at the other side. But if there is no wind at all, or the thatcher is working single-handed, he leaves the ropes coiled at the top till he can return to them, when working back along the other side (Illus. 12).

The 'finish' of the end of the stack depends upon local fashion. You seldom find a perfectly square gable end ; much more commonly the top part of the strakes are continued around it. Sometimes, as a variation of this, they are brought down and secured into the gable end, a little above the eave-level, giving a V-shaped figuring up the gable. The ridge is finished first by bringing the two top trusses together and then bending other drawn straw over from the butt ends to the ear ends, as the 'lie' of the straw passes from one side of the stack to the other (see p. 61).

In Northern localities, or near the coast, where gales have to be reckoned with, 'thrown' roping is used. For this, the rope is wound in stout coils, each about the size and somewhat about the shape of a 'rugger' ball. Two workers stand either side of the stack and the coil is thrown across, backwards and forwards ; sometimes very intricate diamond patterning is achieved, especially on rounded stacks. The ropes are temporarily pegged until the netting is even and taut, and then lumps of rock are secured, working end to end and from one side to the other *alternatively*, for if the rocks were put all at

64

one end they would drag the whole top off. It is surprising how much of the distributed weight even straw rope will sustain, if pegged securely and not allowed to blow about; but strong hemp rope must be used for sea-coast and island stacks.

The straw birds and ornaments which are still made in England as 'finishes' to the stacks are sometimes very delicate and ingenious examples of design. The bird looks best when constructed of long-eared barley; then his tail, wing-flanges, and crest of fringed gold are trimmed to a fine finish, turning him into a Bird of Paradise when the sun shines on him after rain. The diagram shows the two handfuls of straw doubled back to form the fattened body around the butts of the two wing pieces. A little pulling out of the straw for his neck and breast gives him a bold front, with wooden beak and eye complete; as whimsical and lovable a piece of craftsmanship as any in England. When he is made to turn about, he is built over a small dip tin, or even the hollow of a broken plough-handle, and the inserted stick lets him swing in the wind like a proper weathercock (Illus. 23).

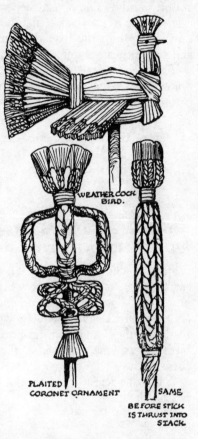

WEATHERCOCK BIRD.

PLAITED CORONET ORNAMENT

SAME BEFORE STICK IS THRUST INTO STACK.

The sketch shows a rick ornament I copied near Usk: I think the thatcher must have helped the stableman plait the cart-horse tails for show days, for he had got the straws even to a hair. The ears were secured evenly around the top of

65

a grooved stick : the straws were then divided into four equal parts and each part plaited separately down to the lower joint. They were then secured again flat along the stick, and subdivided into eight : these eight were then plaited, crossed over with each other, and again secured, the finished ends being cut off evenly. The object then was long and straight, and he carried it up to the stack top. Then, as he thrust the long stake down deep into the hay, the loose plaits thrust up and out, and the intricate symmetrical ornament was achieved.

The men find all manner of excuses for these ornaments. Some thatchers, who work over a wide district, keep a different pattern for each farm and say it is to show 'which stack

A HANDCOCK IN SCOTLAND......

belongs to who '. Others say it is to show ' who t'was thatched stack ', some say it is to show ' they'd more time nor sense ', others say ' corn bird steals no corn and frits off corn-buntin' ' (terrible bird thieves, who in winter dig down through the thatch till they can get at the corn ears within, and show other birds the way in), others say ' they're always done it ', others, cornered, say ' it's proper '.

And ' one reason is as good as another and you'd best rest satisfied '.

§ 5. *Field Stacks*

Stacks vary in shape ; they may be square, oblong, or round. English stacks are usually rectangular or cob. Round stacks are made in Scotland, Wales (and most parts of Ireland). The Scotch have an altogether different way of making hay from ours. At night the hay is gathered into huge handcocks, and when they are ready for stacking they are drawn across

the field by a rope and horse, or pushed on to a small tray
sledge, and one man walks behind steadying the hay on the tray.
Rows of round stacks are set up in the barnyard instead of
one large haystack in the field. Round stacks are usually built
over 'centres' (see
Wood, p. 29), whose
chief use is to venti-
late the centre of the
stack and thus prevent
heating. It takes years
of experience to cal-
culate the exact mo-
ment when grass is
full, and yet has not
begun to wither and
become woody. As
a general rule, the
sooner it is dried and
got in the better. If
rain falls on cut hay,
the water lies across
the grass and half re-
vives it, so that it takes
much longer to dry

A SCOTCH STACH....

than grass which is only moist with its own sap. Nowadays in
large meadows the mechanical cutters and tossers and pick-ups
sweep round, and the hay is whipped up off the field and into
the stack very quickly; but in uneven mountain country,
where small stacks are the rule, heating in the stack has to be
guarded against.

The rectangular stack varies, of course, with the crop.
Material likely to become heated would be built in smaller
and narrower stacks than would dry corn, that would probably
be threshed out soon. Sometimes rectangular stacks have high
pointed tops with a straight ridge-pole continuing full length
to the end. Sometimes the ends are gabled off, like an Essex
roof. Some have low rounded tops: these are called 'cobs'
(and look like it), others 'jugs'—all stacks have local names.

All stacks built by the men in the field are from about

67

6 feet high at the eaves up to 14 to 18 feet in the middle because this is as high as a man can reach with a 6 ft. pitchfork. Now they have mechanical lifts, you can build the stack as high as you like, but it is advisable not to raise the height of the stack past suitable compression weight, for hay is usually sold in compressed bales. Also if there is the slightest unevenness in building the stack that unevenness is going to multiply with the height, and as all stacks ' settle ' considerably, a tilt which can be shored up in a low stack will topple a big one sideways, and a fallen stack again rebuilt is never so good as a straight built one, for the stranded texture of the hay suffers. But do not think that all the poles leaning against stacks apparently shoring them up are propped for uneven settling : sometimes a stack is shored up like this to retard the settling and let the air blow through. A common haylift is made of a single tall spar about twice the height of the finished rick, secured upright with rope stays. To the top is fastened a block pulley taking a running line. The ' crab ' or grip for the hay is at one end of this line which goes up and through the pulley and the other end is attached to a horse. As the horse walks away, he pulls up a truss of hay in the crab. When it is as high as the men on top who are making the rick they yell, the horse stops, they reach out and swing the crab on to the stack, release the hay, and drop the crab over the side. The horse backs in towards the mast, the worker on the ground refills the crab, gives a grunt, and the horse walks out again. Backwards and forwards he goes, and up and down goes the crab— and a pretty monotonous job it is, as I can testify, having led that horse by the hour. There is usually a second pulley fitted about four feet from the ground, swivelling on the lift-pole so that the horse may draw the rope level from his trace. Once, when the rope was taut between horse and pole, in the little pause while the crab swung overhead, a farmyard robin flew down and sat on the line. He was so astonished when the line backed towards the stack that he forgot to let go and rode about four yards before he flew up in a great flurry, looking just like a countrywoman caught on a moving staircase for the first time.

Near Cerne Abbas I found one haylift that had been an old

ship's mast. The newer automatic lift that works like an endless platform is quicker but not so interesting.

Once the stack is built up everybody comes off the top, except the last man. He shapes it, pulling the last loads well in towards the middle and working the top upwards to the shape it will be when finished. When his work is done, he slides down over the end where he will not spoil the stack. If the cart is gone, they mostly leave a truss for him to land soft on. A stack is not often thatched with its own material, as straw is better and cheaper.

As soon as the stack has settled, the thatcher should come. But, when the thatcher comes he always says that they've left the stack the wrong shape—and he fidgets around till he gets it to his liking, though it is not ' thatchers' work ' and he says so while he is doing it.

§ 6. *A Round Stack Thatched with Green Rush*

This small green mountain rush is the medieval ' floor rush '. It is probable, both from historical evidence, and from the evident facility with which country people still plait these mats, that often the floor rushes were merely rush mats and ' laying down new rush ' meant fresh rush matting laid down for the season.

The green rush is used in the West to thatch stacks, and this is the oldest thatch of all, far older than straw. Of old, innumerable things were made of rush in England and rushes may be said to have been plaited into our lives for centuries. Many of the earliest English accounts give payments to rush gatherers ; in one of the old English Romances, the hero is carried up into his lady's bower concealed in a basket of floor rushes. These baskets would be the big two-man-size baskets that are shown in the harvest fields of the tenth and eleventh centuries, not creels which are for one person's back but double-handled carriers swung on a pole on the shoulders of two men. Thus they would bring down the rushes to the castle. In the poem, the lover rises from the rushes and the startled girl cries out ; but when the maidens come running in, she says, ' Oh, it was an otter that sprang against my breast . . .

an otter that must have come from the river hidden in the wet rushes.'

In those days, up in the women's room the new baby arrived on the golden straw of the corn, but as soon as the son and heir could crawl, he would reach down and clutch the strong green rushes, and they would give him the earth's first welcome to his manhood. Later, rushes and sand would scour his armour for battle and the pewter plates for his feast. Rushes were used for chair seating, for bedding, for the shepherd's shady hat: they were built into the plaster of the walls, they were used to enwrap the soft milk cheese, to pleat and goffer the fine linen veils. Bundles of rush were flung into bogs to make it possible for the pack-horses to flounder across; torches and other lights were made of rushes, and there was a ceremonial rush for the floor of the church.

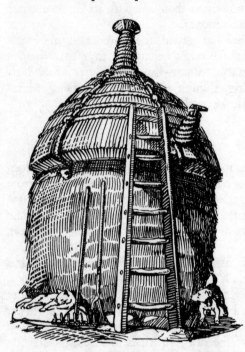

Some of the gypsies can still weave you fancy knots and tokens of white peeled pith, and here and there a countryman will plait strange intricate trifles for you—a tiny rush-bag to hold white pebbles for a child, or queer little nosegays to charm you on a summer afternoon; but the use of rush is gradually dying out. For few things only are still made of rush.

The illustration is of an ordinary stack in Central Wales.

The rush for thatching is the common green moor-rush. It is cut with a scythe at the end of summer, carted down, and used in its green state. It has remarkable properties as thatch, being extremely strong, so that a comparatively thin layer will fend off the heaviest rain. This makes it particularly suitable for the West mountain districts in which it is used.

Circular stacks are often built over frame or stack-centre. They are seldom more than 10 feet high and can be reached by one man with a short ladder. The rushes are laid upwards to the apex (as in straw work). A stout stake is then driven in, a loop of twisted rush rope brought round it, and the two ends secured on either side of the stack. Another crossing of the rush rope is set at right angles to this, again encircling the central stake, and, usually two more, thus quartering the stack according to its size. The thatcher then begins at the top of the stack, winding a truss of rushes around the stake and binding them to it with a close spiral of rush rope. Then he comes downwards, working in a spider's web around and around the stack, looping over each crossing in turn. Sometimes he works spirally, but more often makes an extra ' odd ' line down by knotting into the working line, once in each circle (see diagram). The circles being then complete down to the eave of the stack, he secures the end of the weaving rope, goes round once more driving in all the cross pegs, and the stack is complete.

THATCH MATS. These, sometimes a foot thick, are woven in one piece like a carpet the shape of the roof which is to be covered with thatch. They are fastened on to the underside of the roof before the outer thatching is done. Sometimes they are visible within the house, forming the ceiling of the attic, but more often this mat is invisible as the roof is plastered over below it. I found a perfect specimen taken from a house in the Gower Peninsular which was being renovated and its ceiling raised to comply with the new Government Regulations. This thatch mat was probably eighty to a hundred years old, and absolutely perfect. It was made something like the truckle bed, but with a clever complicated plaited stitch.

§ 7. *Straw Work*

Straw work is dying out in most parts of England, but certain things are still made with it ; and curiously enough the old thick straw work designs are now being reproduced by Arts and Crafts people in raffia. It is exasperating that the use of imported raffia should be encouraged while natural straw has been abandoned.

In wrapped straw work, the straws are ' drawn ' till they lie smooth and level. Sometimes they are lightly twisted into a rope to facilitate wrapping, but more often used straight. The longest and strongest straws are used in sewing and bound with

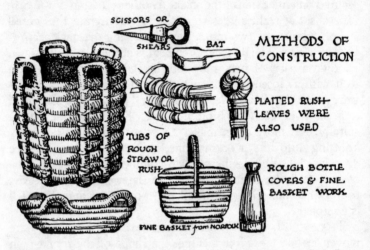

SCISSORS OR SHEARS
BAT
METHODS OF CONSTRUCTION
PLAITED RUSH-LEAVES WERE ALSO USED
TUBS OF ROUGH STRAW OR RUSH
ROUGH BOTTLE COVERS & FINE BASKET WORK
FINE BASKET *from* NORFOLK

a stitching that goes over and over and through. It is very simple work. A rope of half a dozen straws, closely wrapped, would be fine enough for small objects ; for larger, such as chairs, a rope as thick as one's fist is required. This heavier work is sometimes reinforced with the green, pliant strippings from the rod (see Osiers, page 88), because they are longer and stronger than straw.

In the basket illustrated, which was made in Norfolk, green withy was used to bind the handles on to the cylinder of straw, and as the line of osier was continued upwards, it also strengthened the upward lift. Chairs made of straw are still to be found in many cottages, and a few treasured specimens in country houses. The age of some of them proves the strength and endurance of the material. With those that have been well made, the straw tends to pack together in a felted thickness, and the polished surface of smooth outer straw takes on a mellow golden sheen. Straw chairs are remarkably warm : old folk, used to sitting in them by the fireside, complain very much if they have to change to any other.

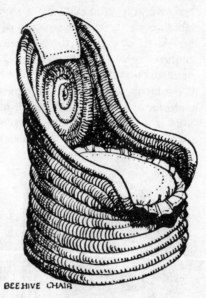

BEEHIVE CHAIR

The diagram shows a hooded form of straw chair, which was sketched at a farmhouse near Brecon. I have found other patterns in the New Forest and the Forest of Dean. This type of straw basketry is known as 'beehive' work, because straw beehives or skelps were made of it, and in the past when straw beehives were in

These chairs are sometimes bound with blackberry bark stripping : they are then harder and a pleasant brownish-green colour. In Wales they are called 'harp' chairs—the hood is about the height of the large Welsh pedal harp. Some modern seaside resorts use for the beach a variety of these chairs which, characteristically, are now made abroad.

common use, beehive makers were probably the greatest adepts. Though skelp-making is still the most representative 'straw' industry we have left, the occasional making of other objects in straw (such as the things hastily put together on a farm) justifies their mention here. The truckle bed, roughly contrived by a shepherd to-day for use in his lambing shed, is a descendant of the skilfully made truckle bed that was part of the ordinary bedroom furniture a century ago. These straw truckles (or possibly trundles—because they *trundled* under the bed) were particularly convenient at inns or in guest rooms as beds for the children or servants. Before childbirth, the expectant mother used it; while the nurse warmed the bed for mother and baby. Hence 'the lady in the straw' did not mean a truss of straw, but the *straw bed*.

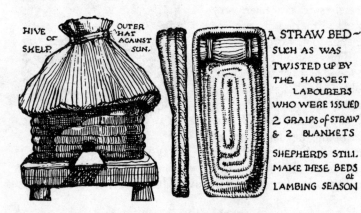

HIVE or SKELP.

OUTER HAT AGAINST SUN.

A STRAW BED ~ SUCH AS WAS TWISTED UP BY THE HARVEST LABOURERS WHO WERE ISSUED 2 GRAIPS of STRAW & 2 BLANKETS

SHEPHERDS STILL MAKE THESE BEDS at LAMBING SEASON

Nowadays in remote country districts harvest work is still arranged for by an influx of temporary workers. These extra labourers are roughly, but comfortably, accommodated in the barn. Stephens in his book *The Farm* gives as details of such arrangements for the men, the issuing of blankets, a sheet, and '*two trusses of straw each to make their bed*'; and he tells us, too, that they were given time to make these beds properly. Straw was not merely thrown down in a pile, but constructed carefully, and the beds, in use for several weeks, were reasonably comfortable. A dalesman shepherd showed me, long ago, how to make one. The straw was roughly twisted out into a long soft rope and then the worker knelt with knee and foot on the first length of the rope and quickly, skilfully passed the rope round behind his foot in front of his knee, back behind his foot and round again, letting the rope lie smoothly on the floor till he was kneeling in the middle of an oblong mat. To increase the length he took a bent-back turn before him once or twice, according to the length of bed he required. Half a dozen long, smooth pointed withies (or spics) were then driven through from side to side, pinning the rope flat and level upon the floor. The bottom of the bed being thus completed, he took a length of twine and a stack needle, and, still kneeling on the bed to hold it steady, sewed the last round or so of the straw on top of the edge of the bed, making a little rim or tray edge all round. This would keep out the draughts and tuck in the blanket. He finished my specimen with a " reet reared head " of two extra coils. The remainder of the straw was drawn lengthways and secured loosely at both ends, and trimmed to make a neat pillow to fit.

Here is an old description of this bed :

" Gromes puleth slyn fyle (fine straw) and make litere
   Six feet on length without disware,
   Seven feet it shall be broad.
   Well watered and wrythen, twisted and craftily y trod.
   Wyspes drawn out at fete and syde. . . . "

I also reproduce an illustration of a small and delicately made basket which came from near Covehithe in East Anglia. It is extraordinarily fine and light, and though of the traditional English pattern, has foreign delicacy of finish. There are too many of these still to be found for it to be possible that all of them were the work of French prisoners of war. But the French straw-workers, who use their material imaginatively, may have been the originators of this finer type of work.

Straw hassocks are chiefly made for cathedrals and churches. The order may be booked through an urban ecclesiastical firm and the finished hassocks be considered as a commercialized urban product, but the thick-twisted straw foundations are usually made in the country. A few years ago, down by the river, I came across an encampment of gypsies, busily completing an order for straw hassocks. ' Oh,' they said unctuously, ' we are all become church workers, yes, indeed,'

and in evidence of hastily acquired virtue, they spread
across a gaudy cart-wheel, a symbol of red felt labelled
Kneel To Pray.

Straw Rope Making has been described (see page 63),
but it is often necessary to make a longer and better finished
rope of straw. The simplest implement used is a swing hook.
There are other types, such as the wimbrel, rather like the
spindle of a spinning wheel in principle, but as I have never
seen one of these in use, I describe the more ordinary
throw-hook. One of the throw-hooks in the drawing was in

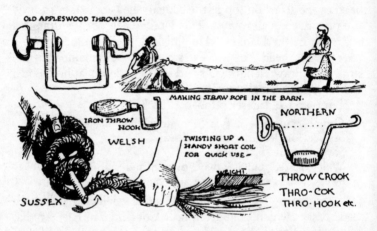

OLD APPLESWOOD THROW-HOOK.

MAKING STRAW ROPE IN THE BARN.

IRON THROW HOOK

WELSH

TWISTING UP A HANDY SHORT COIL FOR QUICK USE—

NORTHERN

THROW CROOK
THRO-COK
THRO-HOOK etc.

SUSSEX.

WEIGHT

use near Llangefni. It was carved from one piece of apple wood.
The one below, where the hook is of iron, was drawn in
Leicestershire. I have seen larger, round curved hooks made
in the North, and even a sickle used as an emergency hook.

Straw rope is usually made on a wet day, in the barn.
A quantity of drawn straw is collected at one end, and the
worker sits among it on a milking stool, his assistant standing
in front of him holding the hook. A loop of the straw is bent
round the hook, which is then given a twist; the worker adds
more straw and the hook is given another twist, till the bent
ends of the straw are safely embedded and the rope is well
begun. It then continues rapidly, the assistant walking backwards continuously turning the hook and the worker feeding

the twisting rope-end with fresh straw. By the time the helper reaches the furthest end of the barn, the straw rope is twisting pretty quickly. A few extra twists are given to finish it and then, as the assistant walks slowly forward, the worker gathers the coiling rope up between elbow and crook of thumb, and the spring of the straw holds the coiled rope in a neat cob shape. Sometimes the worker prefers to wrap the rope round and round sticks, instead of coiling it into a ball. He must clear the floor before starting, as the swinging rope sometimes sweeps it, and needs all the space it can get.

In the North, straw rope-making was thought rather a good opportunity for courting, because if the lass twisted the rope for you, you were given a chance to think what to say to her and she could not get off till you had said it. But straw rope is less used nowadays for stacks, binder-string being nearly always substituted ; though a farmer will sometimes ' throw ' a very thick, soft straw-rope to help hold newly dropped calves ·or foals.

§ 8. *Heather Besoms*

Heather, by which it must be understood that I mean Ling (*Caluna culgaris*), is also used for house thatching, especially in the North, but as I have dealt with straw and reed thatch at length, here I will only describe the heather-besoms.

Besoms are made variously from ling, birch, or marrum grass. The binding or fastening is sometimes of iron wire, but withy, or split ash binding, is considered better, as it is less liable to cut the bristles, and it gives a wider grip. We find besom-making carried on in localities where there are a good many farms and adjacent market towns to provide a reasonable sale. Here the besom-maker will sell a fair quantity, and though the price is low, the material is not expensive and the job can be done at times when it is not possible to work on the land. The besom in the illustration was made in a stone shed in a small village in the Yorkshire Dales. This is how the work is done.

In April, or early spring, the ling is cut : the time must be chosen between 12th August and game bird nesting in the

77

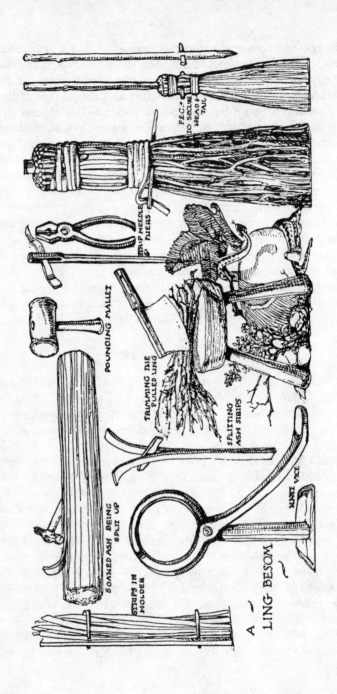

PEG
TO SECURE
HEAD &
TAIL

STRIP NEEDLE
& PLIERS

POUNDING MALLET

TRIMMING THE
PULLED LING

SPLITTING
ASH STRIPS

SOAKED ASH BEING
SPLIT UP

KNEE VICE

STRIPS IN
HOLDER

A ~
LING BESOM ~

spring. The maker selects a good patch of heather with some care. Where it has been burnt very close, for grazing, it will be too short; in hollows which have escaped the fire, where the ling has grown between rocks, it is too high, and the stems will be too lanky. The ling must be 'in sap', that is, pliant. If you bend a handful of ling and it snaps, then it will snap when it becomes a broom, and deposit more on your floor than it should take off it. If you take the heather before it 'sheds', then after it withers, all the banging and beating in the world will never stop it leaving powdery dust as it sweeps. As you buy your patch by the piece, you choose one which has not much bracken or stone among it. You must find a place where the ling has grown straight and clean with fair branching, usually on a level stretch of moor. Moors are not level till you get to the top, so you may have to go quite a long way to find a patch of ling that is worth cutting.

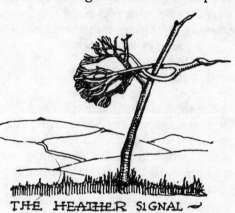

THE HEATHER SIGNAL

I went one day to watch the worker cutting. He used a short straight-handled bille. He had brought a long-legged wooden stool, which straddled across the rough heather and gave him a steady chopping-block. He pulled the ling in handfuls, chopped off the curling roots, very slightly trimmed the ends, and lightly bundled the cut material as he worked. He had driven in a couple of stakes, between which to pile his stack. When he had cut all he wanted, we returned down the moor track to the high road. Where the track came out on to the high road, crossing the moor, he drove in a stick and shoved a twist of heather sideways through the top of it—a signal to the lorryman, with whom he had an arrangement to cart the ling the next time the lorry was passing, that it was ready for him. A few weeks later, when I went to see him, the heather had arrived, and was piled up in the shed; the besom-maker had gone down to the brook that ran through the farm pasture to carry up the log of ash wood which he had put to soak there a few months before. Whenever he saw a suitable piece of this ash log, he marked it in his mind's eye (for it is sometimes exasperatingly difficult to find a suitable piece when you want it). Few town workers realize the enormous number of different things a countryman must carry 'in his mind's eye'. His job involves so many varied contacts, small suggestions, and indications that have a meaning for him. The townsman laughs at the countryman for spending so much of his time 'looking over the gate'; he does not realize how much the countryman is seeing as he gazes over that gate.

This piece of ash, which was now carried up to the shed, was set to drain across the chopping block. It was about 4 feet long and a clean pearly white. It was splitting evenly into lengths, each split the thinness of the ring of a year's growth in depth, and the length of the log. As the wood dried (a process that took days) these strips kept splitting up, and my friend loosed them with a few shrewd blows from a wooden mallet, which freed the concentric circles without bruising the fibres. All 'split wood' for binding and weaving work is obtained like this (see trugs, slops, etc.).

How is the wood made to split like this? Simply by soaking it. The waterproof bark is removed, and the log is wedged under a few stones to keep it down. Swift running water, running the way of the fibres, is best; and you should turn it sometimes when you're passing.

When I came a few weeks later, a good supply of strips were collected into a holder made of a twist of wire and a plank. I have seen a worn thin horseshoe driven into a plank used for a holder, or two bucket handles will do, hooked on to the wall. Besoms are sold wholesale in bales separate from their tails, but a besom-maker usually has a few tails in hand for retail stock. Tails are ash, beech, or any other sapling or stake he may acquire. Occasionally he buys a piece of timber which will cleave into a number of stakes for tails.

The material now being ready, the besom-maker brought

out his iron grip. This was a simple vice with a circular jaw, secured upright into a block of wood. The loose grip-pin was bent outwards, to come about the height of his knee. Sometimes these grips are made to be operated by the foot, but this, necessitating a longer shaft, makes the vice heavier in work. A bundle of selected heather was pulled from the stack and laid in the open circle of the vice : there were a few springing bouncing movements, and then a firm grip, as the long strands of the ling became packed together. The grip held them while the worker with his two free hands bent the ash strip round into a tight binding above and below the grip. He then shifted the broom along, gripped it in the vice again above, and again below, each time binding either side the gripped place. This gave four grips and four *tight* binds. The last bind he put on lower and more loosely, to allow for the spreading of the ling. The besom was then thrown out of the grip and the next made. In binding, he used a strong needle, about 18 inches long, and a pair of pliers to grip the end of the tie. The bind, or tie, varies with each worker : this one finished his tie by drawing the end entirely through the besom, over one strand of the tie and back out at the other side. He ' reckoned that made a good firm finish '.

When I came next day, several besoms had been made, so instead of the grip, he brought out the wooden stool and chopper and squared the ends. The ' wholesale ' besoms, without tails, were then finished. They were packed like sardines, into an oblong block secured with a strand of wire, using the pliers. In packing, a few strong pieces of ling are laid under the wire binding to prevent cutting.

' Retail ' besoms complete with tails have the tails inserted by banging the tail-end down on a stone, so that the pointed end pierces the besom head above it. Another method is for the worker to stand on a wall or step holding the besom clear of the ground, and hammer on the end of the stake. The besom head then jerks up its tail. When the stake has gone in a firm distance, he drives a peg or strong iron nail through besom and tail; to secure it.

A good besom costs sixpence, a superlative one, ninepence with the tail, and they will last a very long time. But they should not be left standing on their heads as that bends the ling: they should be up-ended on their tails, especially when damp. It does not matter if a little water runs down into the handle, it only swells it in tighter.

If the stacked ling is left on the moorside for any length of time, the wild things make free of it. My besom-maker once found a long brown adder stacked into his ling. It must have passed through three lots of hands, while they were unloading and stacking into the shed, but it did not show itself till a week later when he himself was in the shed, with the door open, and saw it slide out of a bundle over the doorstep before he could slat it. It got into the rough by the dyke, where he lost it. He wasn't sorry. The keepers say "adders take eggs," but do not seem to realize that they seldom need to, as plenty of frogs and other small animals live on the moor. We found an adder once that was all bedaubed with honey, as if it had been in the honey of a bee's nest; it was the leaves sticking to it we saw moving, before we saw the adder.

§ 9. *Marrum Grass*

Probably long before the Romans came things were made in England from marrum grass. It grew around the very earliest British settlements and its enduring quality and pliant length would make it a very adaptable weaving material. The marrum grass mats which lay over the earth floors of the early British dwellings, or formed partitions and wind-screens hung within the huts, were probably the exact counterparts of the marrum grass mats made to-day. Marrum grass or rush matting is often carved below recumbent effigies in Churches, symbolical of man's return to earth. These stone mats are identical in design with those now made in Anglesey. I have found the identical design of mats and the remains of baskets and stools made from marrum grass in the dusty crypts of old churches on the east coast of Ireland. One very beautiful specimen of marrum stool I hauled out from unexplored recesses below a layer of skulls at Lusk in Ireland. Reasoning from the date when the church alterations were made, I reckoned that specimen of grass must have been at least 80, and more probably, 120 years old. It was

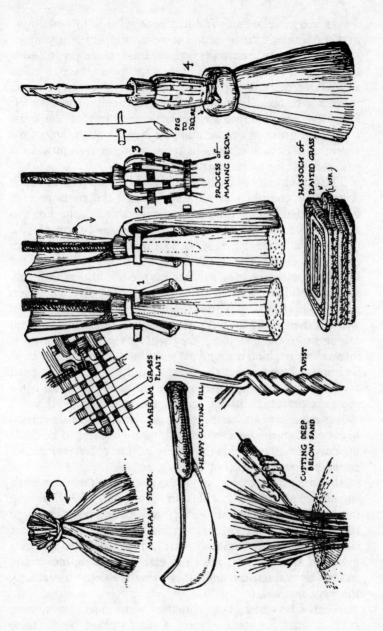

PROCESS of MAKING BESOM

PEG TO SECURE

HASSOCK of PLAITED GRASS "LISK."

MARRAM GRASS PLAIT

MARRAM STOOK.

HEAVY CUTTING BILL

TWIST

CUTTING DEEP BELOW SAND

woven in a circular form and had been filled with peat moss, or some similar substance, and sewn up to form a hassock or stool. By its position in the depths of the ruined crypt, under stones, dust, and odd bones, it had probably been used as a basket to carry down rubbish when the place was cleared. It must have been damped and dried repeatedly, and then left forgotten, yet the grass was as elastic and firm as the grass I had seen growing on the dunes at Newborough, across the water. Under such conditions straw, reed, or rush must have perished, but the marrum grass endured.

The grass is cut late in the summer. The deep sand in which it grows is continuously shifting, so that the roots become deeply buried and the grass is of considerable length. For this reason the cutters use a coarse, heavily headed form of bille (p. 83), usually made for the purpose by the local smith. With this they cut the grass well down below the surface of the sand, gathering the grey swathes on the left arm as they fall. As an armful is gathered, it is carried to one side where, on a level spot, the marrum grass is set out to ' win '. The women who do the cutting are very skilled in the quick binding and setting up of these stooks ; they whip a few strands of grass into a short tie, bind it round the grass bundle, and with a swift whirling movement set the grass upright in a steady green cone. It is a knack, for at this stage the newly cut grass is the most exasperating, slithery, limp, and unmanageable ' thrave '. When enough grass has been cut and dried out, it is carried home and stored in the small stone outhouses, to be made up during winter evenings. Many of the older women are incredibly quick workers (Illus. 25, 26).

Mats are made in many ways, the commonest method being the plaited strip mat. For this, the grasses are plaited in long flat strips. These strips are rolled into balls and flung back into the shed till a convenient time to make the mats. This is a more cumbersome job, requiring space, the strips being sewn together, side by side, until the mat is the required width. Several central African and other primitive tribes use exactly the same method.

When I was there last year, they were making a number of 7 ft. mats for stack covers. Crops stacked below these marrum grass mats will never heat; for they fend off quite heavy rain, but are much more open and therefore cooler in use than a straw thatch. They last several seasons, if well stored. This year they are, I believe, to experiment with narrower strips for laying down between strawberry rows. The cleanliness and resilience of the marrum grass mats should make them particularly suitable for this, as they can be trodden on repeatedly, and yet not become soggy. I have used them myself for storing fruit, nailing them to the wooden frames of the apple trays over open laths. The open weaving lets the air circulate freely, while at the same time the closer packing keeps out the frost.

Brooms and brushes are made from the marrum grass. Most of the coastal houses near the sand dunes are thickly limewashed, and this whitening is brushed on with marrum grass brushes, or rather short hand brooms. In the simplest form, an oblong hole is burnt or cut through a slat of wood, and a double handful of grass thrust through from either side, so that there are equal number of butts and tips mixed on both sides of the slat. The grass is then bent down and tied beyond the end of the slat, which fans it out and turns it into a very serviceable whitewash brush (see illustration). Last time I was in Ireland, I noticed the same pattern on sale in small country shops.

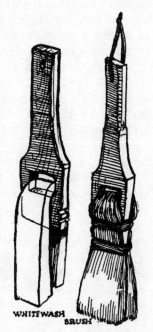

WHITEWASH BRUSH

The broom proper is drawn in some detail on page 83, as an excellent example of the very decorative result that can be obtained from conditions that are severely utilitarian. Two thraves of the grass are securely bound near the end of the broom-stick; the first thrave has its butts downwards to form the stiffer heart of the broom, the second thrave has its butts slightly overlapping the tapering tops of the first thrave, and its soft tips up towards the handle. The first binding is made secure. The upper thrave is then turned down over the binding, exposing the tips of the first thrave and covering the butts at their lower end. This gives a soft dust cover to the stiffer centre of the broom. Sometimes this process is repeated twice, which gives a more bushy end, but for ordinary purposes the double thrave gives an adequate broom. As the upper thrave is turned down, the binding cord is woven in and out the strands, each in turn. It can be more easily seen in the illustration than described; the separate binding of the strands gives a tight hold to the grass and a very decorative finish. Where the outer grass passes over the extra thickness of its butts, another binding is passed round and drawn very tightly. Above this a nail is usually driven through the wood, as added security. A loop of string to hang the broom up so that the dust may fall out, and the fibre be straightened after use—and you have as neat a little broom as any made in England.

§ 10. *Fine Work in Rush and Grass*

People who see only the rough rush and grass work of to-day do not realize the exquisitely fine finish and delicacy of design of which the older rush workers were capable.

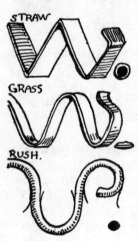

STRAW

GRASS

RUSH.

As a designer myself, I am convinced that much intricate Celtic strap work such as the Book of Kells, or the bolder designs on some of the Stone Crosses, were based on the patterns used in green rush or sea-grass work. These green and pliant materials would be much more common at a period when corn was scarce and straw kept for fodder; frequently in those days the grass was allowed to grow up among the long stubble and mowed together with it. In the natural weave of rush and sea-grass, the turns, though

86

abrupt, are slightly rounded, and while green the material follows a very constant curve. Straw is hollow and cylindrical, and so will always flatten in plait and crack in turning, if the angle is sharp. Thus we find a marked contrast between the angled English straw plait and the curving Celtic scroll.

In early English translations of the Bible, Moses is floated in a basket of reed, and reed and rush work is mentioned repeatedly in all parts of England. As late as 1700, references are made to 'neat rush boxes and baskets, nicely executed'.

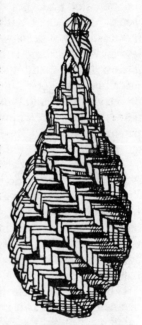

In my own memory the gypsies used to make knots and 'fancies' of white peeled rush, intricate patterns as finely executed as ivory carvings. Country lads at hiring-fairs used to make cockades for their hats, and button-holes of polished gold straws, and sometimes a thatcher will plait up a 'nek' of fancy straw weaving.

A couple of years ago, passing through Anglesey, I stopped at a stone bridge, on which sat a post-man, awaiting his mail van. I don't know what impulse made me suddenly turn and ask him if he could 'plait little toys of rushes'. He looked surprised—but without saying a word, he bent over the little bridge, to where a small stream fingered its way between reedy shallows. Silently he selected a handful of reeds, and sitting there on the stone wall in the autumn sun-shine, he plaited me a perfect small

DELICATELY WOVEN RUSH TOY. (ANGLESEY.)

green cage or rattle box, in which two small white pebbles clicked together for 'birds'. It was 4 inches long, perfectly pliant and elastic in weave, taut and strong and delicate, *exactly like* the one given to me as a child that had been in my mind, when I asked him.

§ 11. *Willow : the Osier*

" The Osiar commonly groweth of his own self, and is also
planted of his roddes, in watrie and marish grounds ; it is planted
and springs most plentifully, where the earth is beaten up with the
rage and overflowings of the water, and it serveth as a sure defence
for making bankes and walles of marshes, and that chiefly in
Marche.

" Osiar serveth for making baskets, chayres, hampers and
other countrey stuffe."

Gervase Markham : circa 1620

The osier trade has declined by half during the last ten
years.   Competent authorities attribute this decline to the
increased cost of keeping large tracts of land drained, although
less and less acreage is being cropped. Also the home trade
is now in competition with the large quantities of foreign
osier and lias goods imported. But the usual reproach of
country conservation in design is undeserved here for the
new designs for willow furniture are sponsored by competent artists, and modern willow work is as pleasant in
appearance as it is comfortable and durable in use.

The decline in demand seems to date back from the Middle
Ages, though the trade has known ups as well as downs. In
Victorian times there was a demand for elaborate basket chairs
for drawing rooms and there is a demand to-day for plain
basket chairs for use in gardens. During the Great War
thousands of woven osier shell-cases were needed, and certain
agricultural experiments are now calling for special willow
hampers.

But in earlier centuries, willows were probably the most
necessary crop grown in England. Timber, and the cumbersome means of working it, were beyond the means of poorer
folk ; they got hold of some timber for the structural posts of
their huts, but the rest would be built of willow and clay.
Woven rush and split logs, planks squared out with the side
axe and adze could be used, but in the days of hand sawing
in the pit, wooden furniture was scarce and very heavy.
Willow, useful and light, grew all over the great marshes and
grew quickly, so that, even on mountain land, a small marsh
or a pond by some stream would keep up a good supply of

rods, with little tending save for the annual cutting. Very early the Westminster monks had an osier bed where Westminster Hall now stands, and many old manuscript illustrations show willow in use for every type of building or fencing, whether in the sheep-fold or castle, in peasant's hut in peace-time, or fortress in war. Some twelfth-century manuscripts even depict a bishop sitting in a fine solid-looking basket chair.

For centuries, architects and builders used scaffolding of plaited willow, and carried stone, mortar, and sand in creels and baskets of willow.

AN INITIAL FROM M.S.COTT DOM XII

12ᵀᴴ CENT. BISHOP IN A BASKET CHAIR.

Norden wrote in 1607: 'I have planted an osier hope (for so they call it in Essex) in saturated ground, that before was no use ... and I think it yeldeth me now greater benefit yearly acre for acre than the best wheat.' How much of this 'greater benefit' remains to us to-day? To study how far are we using willow as our ancestors used it we have arranged the uses under the following headings.

THE ANCIENT WATTLE AND DAUB WORK. This has practically died out, but great wattle platforms and fences are built to reclaim land or hold dykes, on the same principle that formerly used willow for encasing in hardening mud. Lately we have had Dutch workers to 'Mattrass' part of the Wash, to reclaim land, making wide 'beds' of willow, towing them out, and sinking them with stones into the sand.

WAGONS. Early wagons and carts had superstructures of wicker work. Light basket 'chaises' lasted in use till the last century; but now an occasional hurdle extension to enlarge some farm cart for the hay or corn harvest carting is all that remains.

PANNIERS OR CREELS. The creel for carrying peats (especially in Ireland) exactly reproduces the thirteenth century earth-carrier creel, but except locally or on the stony islands in the West, none are made new now, except for kelp.

PANNIERS. The long packhorse trails created a need for numbers of pony and donkey panniers made of willow. Nowadays there are panniers for hill ponies, or sea-side cockle diggers, and some panniers for children to ride in, but the demand is small.

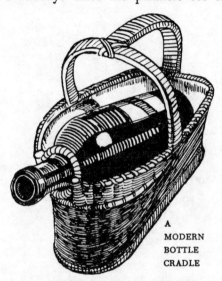

A
MODERN
BOTTLE
CRADLE

MUZZLES. Many of them are still made of willow in Ireland for weaning calves, and a few, too, in isolated places in England.

CORACLES OR WIL-LOW CANOES. The basket framework was probably more closely woven for the early British coracle than for the few made to-day. Some willow is still used, but not much.

FENCING. 'Hurdles' have rather taken the place of the lighter osier-pens for sheep.

FURNITURE. Cradles and light cribs of willow date from before Chaucer, and are still in use. Chairs are still made of it, also stools and a few tables (but the latter are structurally seldom satisfactory as willow is the wrong material for anything demanding rigidity).

BASKETS AND HAMPERS. These are still made, and their infinite variety is probably as great as ever. The willow fruit basket fluctuates with the fruit trade, but there is always a reasonable demand for trade hampers and baskets. For example, laundries require square willow hampers, and usually like them to be of white peeled rods. Watercress hampers to withstand the damp are of unpeeled rod. Many firms like circular hampers for dispatch of their ' gift ' and ' Christmas ' goods, as looking more appropriate than the heavier wooden crate.

IN CONNECTION WITH POTTERY AND GLASS. The old medical urine-bottle basket that was part of the indispensable equipment of the seventeenth century doctor, is reproduced in the butler's wine-bottle basket of to-day, both being made and carried with care. The large baskets and creels that carried glass and china along the packhorse routes have ceased to be made, but large jars, demijohns, distilled water, and spirit flasks are still put into basketwork containers.

FOR TRAPS AND HIVES. And the same lobster pots, crab pots, bait ' pads ' and ' scoops ', eel-hives and fish kiddles, bird cages, cat and dog baskets, bird-baskets and bird-houses or ' cotes ', are made of willow to-day as they were made hundreds of years ago.

§ 12. *Preparing Willow Rods*

Willow rods, or ' sallies ' as they still call them in Ireland, are grown roughly and cut somewhat haphazardly by the old countryman who wants to do an odd job, but willow that is used for furniture, and all commercial willow, is grown and cultivated very carefully.

The flat water lands, or ' moors ' as they are called in Somerset, supply a large quantity of the willow of commerce. The drains or ' rhines ' that intersect this low-lying land make it very suitable for willow cultivation, and many special rhines were constructed for this purpose. In the winter it is a wild desolate country, haunted by the cry of marsh birds, the wind sighing in the withered sedge, and the sobbing and choking of dyke water. The rain-clouds from the west trail across the water-logged lands, and the lonely dyke roads are long and solitary beside the land-locked waterways. The very names hold magic : Dilcheat, Sedgemoor, Isle of Brewers, Long Lead, Thorn Falcon, Atheleny, and Old John, near Glastonbury, and Wells. In spring it is perhaps one of the most lovely districts in England; surrounded by the apple blossom of the cider counties, the green and silver moors are lit by the willow buds.

Later the April rain hangs in pearls on the dusty pale gold catkins, on the silver waterways bob joyously fluffy yellow ducklings. They try to get the ducks hatched early, because the elvers (shoals of tiny thread-like eels) come across the Atlantic and up to the long rhines in spring, and the natives of the moors catch them in white nets and cook and eat them. They are considered fine and fattening for babies, as well as for ducklings.

In spring the rods are cut. Men and boys go down the rows with knives and bend the rods dexterously across their knee, and under their left arm-pits, collecting them into a bundle and binding them with one cleverly noosed rod. They lay them down as they finish them and continue down the row. The bundles are then either stacked on the narrow dyke roadway or carried from the moor in shallow double-ended black boats that are pushed along the narrow rhines with poles. Those needed for white work are then up-ended for a period, in piles, either in the dykes or special ' pits ', and stand thus till nearly June, when they can be peeled easily, leaving the willow rod within a clear, clean white (Illus. 24).

The rods required for ' buffs ' are prepared by boiling. All about the countryside there are long tanks—some specially made, some adapted from ships' boilers (which are a convenient shape)—and fires kept up in furnaces under them night and day. The bundles of willow are stacked into the boilers, head to tail like sardines, till they are cooked thoroughly. Growers contend that the ' buffs ' (i.e. these boiled willows) are harder and wear better than white; the boiling would certainly kill any insect or decaying growth that might cause subsequent damage. The buffs, after peeling, are often left in the open air for some weeks before sale and transport, so that they are well hardened.

The peeling of the white rods is done by women and girls, and almost every child in the district, or any old lady owning any sort of shed, will ' strip ' during the season. The pay is not high, but it is work in which skill and 'knack' are as highly valued as strength. Most large growers have their own peeling sheds, and collect their own lorry-loads of

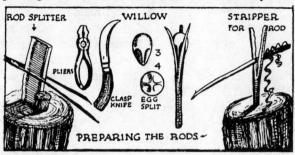

'strippers' as required; it may be said that it takes well timed management to keep cutters, boilers, strippers, and packers working in time with each other through the vicissitudes of an English spring.

The stripping apparatus is simplicity itself: a two-pronged 'fork' (probably nothing more complicated than a strip of blacksmith's hoop iron conveniently split) is jabbed upright into a post. The peeler stands facing it and 'whips' the rod through it—first from one end and then from the other. The thin bark sloughs off and falls behind the fork, and the clean stripped rod is tossed aside and another taken. Rods are sorted into sizes by an equally simple process. They are thrown into a barrel (which is sunk into the ground for

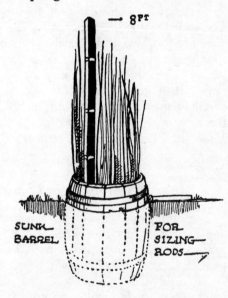

convenience), and the tallest are pulled out first, then the next tallest, and the lowest are bundled together last. Sometimes a measuring rod stands in the barrel or is nailed to the side, with the sizes 8, 7, 6, 5, 4, 3 feet marked on it, and the willows are then ' drawn down ' to the exact size required ; but for general work the sorter carries the size ' in his head ', as well as, literally, at his finger tips. The peeled rods, either buff or white, are bundled and stacked into arm-load measures, variously called ' bolts ' or ' thraves ', and the material is ready for the basket worker. Sometimes a decoration of soft plaited leafwork is used with the willow, and sometimes different coloured rods are introduced, but this ' fancy work ' is going out, and the fashion tends to simple weaving of really good design.

Not all willows are green. For example, some early rods are cut from bright golden yellow or vivid emerald green, because they make the white celery they are tied round in the Vale of Tewkesbury look so snowy white by contrast. The spring-onion growers have a fancy for a neat attractive scarlet rod to bind their bunches of white and green. Some rods are bright gold, red, or deep purple, and many are now being cultivated for their beauty for use as hedges and screens rather than for their utility as basket rods. Varieties of willow grown in Somerset include Black Mould, Champion, Black Spaniard, Osier, Red Bud (a pretty rod this), Dicky Meadows, Black Rod, and Canadian.

§ 13. *Furniture and Basket Willow*

These prepared willow rods are used in all light basketry and for *indoor* furniture work. Garden furniture is usually made of natural unpeeled willow, as being more weather-proof and more in keeping with outdoor surroundings.

Probably the earliest piece of wicker-work that still continues in use is a child's crib or cradle. Early manuscripts show English, Flemish, and Norman-French mothers all carrying their babies in various forms of light wicker frames (the idea was to make their limbs grow straight). All

94

agricultural peoples develop some sort of carrier for the suckling baby during the mother's field-work, the 'baby on the tree top' was probably an actual fact—sensibly hung up there in its light crib while Mother did her communal hoeing. For indoors, the light wicker-work cot was raised to stand by the mother's table or bed. Thus the Jolly Miller's Wife in Chaucer's Reeves tale, when the family go to bed :

> The cradel at hir beddes feet is set
> To rokken, and to yeue the child to souke.

The drawing is a reconstruction from a woodcut of the Basket-makers' Coat of Arms of the sixteenth century. The woodcut itself is thick and coarse, but the general outline seems a very adequate basket-work cradle. The pattern and period of the design make it probable that the body-work of the crib is made of rush leaves, or some soft padded weave (which would be warm and requires less padding); the stakes

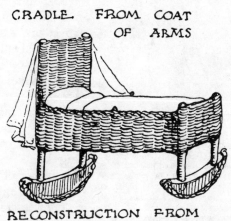

CRADLE FROM COAT OF ARMS

RECONSTRUCTION FROM WOODCUT

are used exactly as they would be to-day, and the rockers are apparently of wood, with an interesting line of soft plait shown *under* the rockers. (This small point is specially intriguing to the author. The plait may be only the artist's design in drawing the Coat of Arms, but an heraldic artist is usually extremely accurate in detail, and it seems more likely to be an intentional padding secured around the wooden curve of the rocker. Now on bare earth flooring (as in a Scotch croft or Irish cabin) or on earth-set stone flags (as in a Yorkshire kitchen or medieval castle), rocking does not

95

make an echoing noise, but my mother tells me she had to have our heavy oak cradle so padded for her first-born because the worn wooden rockers made such a thunder on the wooden floor.

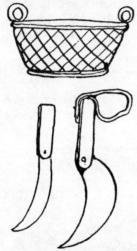

These drawings are taken from a coarse woodcut, interesting only as showing the type of knife. A strong, rather hooked clasp knife is used to-day.

The basket probably conventionalizes an exact replica of a washerwoman's clothes basket. For stage purposes the buck basket of *The Merry Wives of Windsor* is usually lidded; but it was more probably of the above shape, open and deeper, slung from the shoulders of two workers by a pole through the handles. Floor rushes were carted in the same way.

Thus the reconstructed padding of this cradle is an interesting historical point, because at the period of the woodcut, level wooden floorings for upstairs bedrooms were becoming general, and the rush that used to pad the floor was being swept away.

The records of the Basket-makers' Company were all destroyed in the Great Fire of London, but the 'Oath of the Worshipful Company of Basket Makers' (1569) still survives with their Coat of Arms :

'Asure, three cross baskets in pale argent between a prime and an iron on the dexter and a cutting knife and an outsticker on the sinister of the second.

'Crest on a wreath of the colours a cradle therein a child rocked at head by girl, at feet by boy, both vested all proper.

'Motto, "Let us love one another"'.

§ 14. *Green Willow Work*

The green willow is the living tree and will remain alive for a long time in water. Thus all eel-hives, fish kiddles, lobster, and crab pots are made from green willow.

Everyone knows the shape of ordinary lobster and crab pots, they are made in quantities around the coast where such fishing is done (and Leo Walmsley in *Phantom Lobster* has said, once and for all, all there is to say about the inevitability of the design).

There is one interesting detail in the illustration. A small bundle of spare rods is usually sent out with the pots; these are to hang the bait, and you will see that it is well pushed in along the stick so that it hangs close to the inverted opening —to encourage enterprise down the hole, and discourage hooking out through the bars.

Crab pots are like lobster pots, but for crab you weave a lower pot. Willow ' pads ' or open baskets accompany all *a* fishing boats to hold the catch. I have been out overnight with both crab and lobster pots, and no one seeing ' basketry '

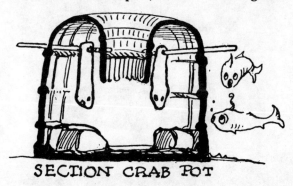

SECTION CRAB POT

in use under salt-water conditions could ever think of wicker work as frail: it is extremely durable and tough.

Kiddles are large basket-work fish traps. Used with fine nets, they did so much damage that in the Middle Ages, when freshwater fish was wanted for food by the country people, and preserved or ' rented ' by landowners, the use of kiddles was forbidden; so that if a kiddle of fish was found illegally in a water-way where it had no business to be, it ' was a *privy* kiddle of fish ' and the consequences were disastrous for its owner.

Kiddles are made and used to-day, at certain times and places. For example, if a mill-race is being cleared, they will ' run out through a kiddle ' to save the fish. They are used, too, on fish farms and in trout hatcheries. A form of kiddle used for eels is called a ' grig '. It is placed in the middle hole of a net, stretched across the narrow part of a dyke or stream, so that the descending eels must, therefore, swim into it and can be lifted out as they come without disturbing the net. A net cod is sometimes substituted for the grig. A ' hive ' is a smaller form of eel trap. These wickerwork fish traps are used in the fens and rivers of the Wash and up the Severn on the West. The Severn type is usually wider, and called a ' will ', and their nets ' will nets '.

The following special example was made by Mr. Killingsworth, of Earith, near St. Neots in the Fen country. He cuts

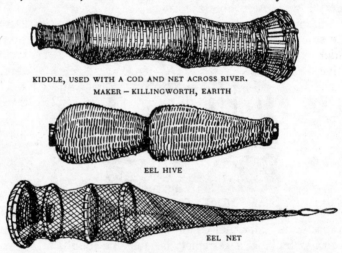

KIDDLE, USED WITH A COD AND NET ACROSS RIVER.
MAKER — KILLINGWORTH, EARITH

EEL HIVE

EEL NET

his own ' rods ' of willow, as convenient, and uses them very fresh. He ' splits ' the weaving rods, using the small egg-shaped wooden splitter shown on page 93. The structural rods are not split. The weaving is begun around a wooden block, which acts as foundation for the weaving, and a gauge for the size. These round wooden cylinders on which the first turns are made are called ' chairs '.

After he has woven around several inches, the long whole rods are bent over and doubled back (see sketch) and the wooden ' chair ' is slipped out (and used again for the next). The weaving goes on steadily, round and round, the maker giving the hive a subtle outward curve.

It is pleasant to watch the quiet skilful worker. He sits

low on a stool, the structural withys 5 feet long waving out in front of him as the hive turns round and round to his weaving. He has that look that comes to all men who work their wits against wild things, of thinking with their fingers. It is a curiously shaped funnel he is making, but its shape has evolved naturally between the flow of running water and the ways of eels—and of fen men. 'Very queer things, are eels,' says the weaver, and the long green withys that twiddle and wave through the door of the shed or whisp along the floor seem to whisper back, 'Yes—very queer things are eels— very strange ways.'

The weaving continues in silence for about a foot, then he stops and peers over his glasses, 'Where's my measuring stick?' Very carefully he measures the length of the curving hive, an inch either way—and 'out goes all eels'. The up-standing, in-turning, ends of the structural withys are left, a little longer than the weaving, like the down-turned 'prongs' of the lobster pot.

I give this process in full detail because people are apt to think rural work simple and easy; it is not easy to adjust those wicker points. Once, watching a Chinese maker of fish traps, I inquired why he turned his work at a certain slant. Language was no use between us, but squatting on the sand, with a twinkle in his eyes, he tapped his forehead, and then commiseratingly indicated the fish, saying in effect, 'I think more cleverly than fish—so I catch fish.'

The English maker of eel hives had just that detached and very thoughtful look as he settled this delicate part of the willow hive. I accused him of 'thinking like an eel'. He smiled quietly, 'I've known them a very long time; they are very queer things, eels.' Then he completed the weaving.

There is a wooden plug thrust into the container end of the trap stopping the hole through which the eels are extracted, and a slip of split willow, on to which the bait is attached, is inserted through the trap end. The hives take about two or three hours to make—well. They last a full season. You cannot use 'preservatives' on freshwater willow, as it would smell in the current, and warn the fish.

The hives are tied to a stone and 'sunk' underneath till

thoroughly water-logged—about nine days. They are then taken out and laid in likely spots. A knowledgeable man will sometimes lift them full twice a night.

This work, with the natural willow, takes more strength than that with the peeled rods. All basket-makers develop beautiful hands, but the workers who make things of green willow become exceptionally strong in the forearm.

§ 15. *Wattle and Daub Work*

Most old castles had to be of timber or stone, but long ago any peasant could build his own wattle and daub

This house, when new, probably had a central fire and roof-hole chimney. The brick chimney of the gable end is obviously built on later.

TYPICAL WATTLE SPACING [BERKSHIRE]

hut. The larger and more permanent buildings would have solid framework, where the 'wattle' would be of hurdle strength and the 'daub' thick, but many frailer huts were probably of much lighter willow, the chinks filled with dried moss and the daub thin. Such would be the short-time hut of a leper, built to last only as long as he lived. We know that, by law, at his death, the leper's hut was burnt to the ground and must have been fired easily, so that it is at least probable that there was more of the inflammable woven wattle work and dried moss in its construction than 'daub'. Even lighter would be the wattle 'arbours' and summer houses and wicker-screened alley ways of the medieval garden, or the tilted wattle screens that caught the flying arrows in the castle under siege.

On the other hand, some of the earliest wattle and daub work was so heavy, strong, and permanent that it stands to-day; this is the true wattle and daub work. Here is an interesting constructional point. In building, a timber framework is first erected (rather like the 'frame' houses of to-day) to support the wattle walls. Where this framework is drilled

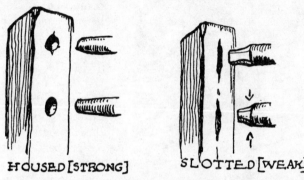

HOUSED [STRONG]   SLOTTED [WEAK]

by auger holes, to receive the ends of the wattle hurdle, it is called 'housed' wattle and lasts well. But where there is an axe slit and the ends of the hurdles are unduly cut down for insertion in these slots, then a weak structural line is formed, and presently, within a century or so, the whole piece falls out. As this happened pretty often, and as 'bricks' became more commonly used, very often original wattle and daub frames were later filled in with brick.

This repair-work must not be confused with the regular timber and brick construction, often so finely designed; but where we find timbered, or half-timbered houses, whose awkwardly shaped wall spaces seem to be brick-filled at random, or where there is much difference in level between the old timber and the brick surface, we may conclude that these awkwardly shaped spaces were *probably* originally filled with wattle and daub. If the bricks are extracted, the *slits* down the sides of the timbers can be seen from where the wattle broke out. The 'housed' wattle lasted better than the slotted.

CONSTRUCTION OF WATTLE & DAUB HURDLE, HOUSED IN TIMBER FRAMEWORK

As the wattle was woven into place, the first packing of daub was worked in. This first layer was *very* firmly driven and packed into the interstices. It was largely composed of neat's hair or goat's hair or, at the worst, straw, reed, or wiry grass. This first filling of 'fibrous plaster' was left as rough as possible on the surface, but the strength of the wall depended on the wattle being thoroughly well 'filled' with this binding layer. Animal hair is practically indestructible; so by the length and colour of the hairs teased out of a piece of broken old wall (of which you know the date) you may deduce to what extent the local breed of cattle has continued or been modified.

When the structural part of the wattle and daub was done, the finishing coats of plaster seemed to ask for scratched and moulded decoration: some of the early houses of wattle and daub, with carved-out timbers, and moulded-on plaster, are perfect examples of delight in ornament.

§ 16. *Hurdles*

Though woven as in basketry, the making of the strong hurdle is a different trade. The hurdle-maker usually works in the open air, and frequently uses other woods than willow; hazel, split ash, and other coppice growth make good hurdles. For a different type, split wood is used.

Historically, hurdles are as old as anything made in England. In a Justian manuscript of the tenth century, the tiny Saxon figures skirmish around hurdles of all sorts. A bear pit has a circular hurdle fencing, and it is taken for granted that Adam used hurdles in his garden.

Under ' Coppice Work ', the preliminary work of the hurdle-maker has been described, for a serious part of the hurdle-maker's job is choosing good material. Hurdles are always of English wood ; I have never found a hurdle-maker

HURDLE MAKING FRAME

AND TOOLS

able to use any stakes but stakes of his own selection and usually of his own cutting.

Hurdles are a southern counties trade, for in parts of the North of England they are practically unknown. This may be due to there being less suitable quick-growing coppice, and also because hurdles are much used for folding sheep. Now you cannot hurdle ' fold ' a wild, black-faced sheep, because it is not convenient to weave a hurdle higher than arm level, that is, 4 or 5 feet, and mountain sheep jump that high easily, so, where a southern shepherd would build a lambing pen of hurdles, the northerner uses stone walls, or a shelter of stakes and cut furze, or ling, bracken, or nets.

The sketch shows an average ' make ' of hurdle : the uprights are driven into the stand, and the worker starts the first weave downwards, obliquely, so that the loose end is subsequently secured, woven over. The top end of the finished weave is worked downwards with the same action. The 'end stakes' are longer; to make stands, or feet, and sharped for thrusting into the ground.

I have sketched a hurdle stand that I saw in use in Gloucestershire, but the stands are commonly squared timber, and no two makers work in quite the same way. When complete, the hurdles are stacked flat under weight (perhaps an old gate and some logs) to ensure flat packing for transport and straight sides.

It may be interesting to add a note on the way a hurdle fence, or enclosure, is set up.

The drawings were made near Selsea Bill, on a bright blowy spring day. The sheep were coming off the Downs that evening at folding time, about 6 o'clock, and one man and his labourer were due to have a newly prepared turnip field ready for their arrival. The lorry bringing the hurdles should have come that morning but did not arrive till afternoon, so the chill spring wind was warmed by a few well-chosen words. (This has nothing to do with the hurdle setting; but I mention it, as it shows there was no time to spare, and so the 'setting' had to be carried out, extra-methodically, and quickly.)

It is noteworthy, therefore, that even though pressed for time, the two shepherds carefully carried one of the new season's hurdles across, and compared it seriously and closely with some of the old last season's hurdles already in the field.

Having decided the new hurdles were up to standard, they directed the lorry man to put them down in two piles. Then that lorryman drove off, and the sunlight that flowed down over the downs shone on two men and two piles of hurdles.

Now consider their problem. You must enclose your turnips, not your sheep, for the more room the sheep have, the better; so the shepherd does some calculating.

You should begin enclosing the field at the driest and most comfortable spot for the sheep, for then the sheep will have the full use of that dry spot from the first day they are in the field. Also, by arrangement, once you have set the complete hurdle line, you should not have to shift more than a couple of hurdles any one day afterwards. The simple countryman does this sort of problem in his head, and thinks nothing of it.

So these two men had another look at the sun, to see how late they were in getting a start, and then they began carrying

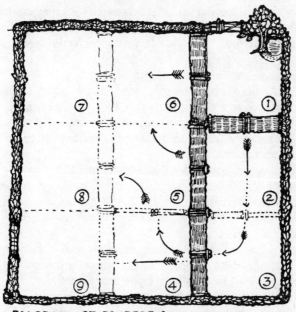

DIAGRAM OF HURDLES IN FIELD
IN THIS CASE TO LAST 9 WEEKS

the hurdles and laying them down along the line where they were to be put up.

When the hurdles were spaced out, they started to 'set' them. One man lifted the hurdle upright on to the line, and thrust one long sharp end stake into the earth. The other shepherd, standing on the far side carrying a mallet, drove down on the two stakes, and the first hurdle was 'up'. Each man gave it an experimental waggle (and both grunted, in a disparaging manner, because they reckoned the hurdle-maker had stinted the length of his end stakes). However, the hurdle stood firmly up for its maker, and they moved on and set the next, and the next, so on down the line, one man lifting and setting, the other man with his mallet on the far side driving the stakes.

After completing a certain distance, they stopped to drive in an extra 'spare' stake, as supplementary buttress to the hurdle wall. (I have seen these supplementary stakes driven at an angle, or rope and 'peg' stakes used.) As they completed one length, they returned down it, tying each hurdle to its neighbour, at top and foot. It was a perfectly systematized job.

The field was just ready, and they were standing back, looking at the completed job, when there came the tonk ! and tonk ! of sheep bells, and the flock of sheep, with their shepherd and old English sheep dog, came in a small grey clamorous dust cloud moving down the lane, to pour into the new enclosure. . . .

Lowland Scotch hurdles are mostly made of wood, split larch or ash. They are more like gates, and one man alone can lift and thrust them into the earth, because he can shove with his foot on the bottom bar, but woven hurdles have no bars.

In hard ground it saves 'racking' the hurdles if you stamp holes for their foot stakes with a slender 'dibber'.

Some woven hurdles are made reinforced by diagonals *nailed* on to the end stakes, but the best hurdle makers I have known condemn this usage and make them plain.

CHAPTER THREE : STONE

§ 1. *Mines and Quarries*

England is a well good land, it were of each land best,
It is set in the end of the World ; afar in the West,
And England is full enough of fruit and of treen—
Of woodes and of parkes, that joy it is to seen.

Of Welles swete, and colde snow, of leafen, and of meade,
Of felner, ore, and of gold, of tyn, and of lead,
Of stele, and of iron, and of bras, and good corn great wonder,
And of the white wool good, betere, n'er, maybe, none other.

ENGLAND HAS always been famed for her minerals, since the
time of the Phœnicians who came for the Cornish tin. The
Romans worked mines in Britain, and Welsh gold is by no
means a new discovery. It was worked by the Romans at
Gogofau Pumpsant, and also occurs in quartz veins west of
Llandovery, in the Arenig rocks, at Clogau and Gwynfynydd.
They say it was once found in small quantities in the alluvium
of the Dolgelly district. Cwmersen gold mine was in hard
flags of slate, and the gold mines of Hafod-y-Morfa are,
I believe, in slate too. According to geologists, the ' numerous
quartz veins intercepting the beds near Dolyfrwyswg all
contain a little gold '. And it is recorded that the total weight
of gold obtained in Merionethshire, 1844–1866, amounted
to 12,800 ounces with an average value of £3 4s. 0d. an
ounce.

It is largely a balance between the price of gold and the
cost of production that makes this native gold worth working,
or not.

Some silver mines in North Wales (silver is found in
connection with other metals) were tried out again only
a few years ago, but phenomenal floods and a series of
accidents destroyed what might have been—after all—only
a disappointing trial (Illus. 27).

It is interesting that England differs from most of Europe

in its definition of what is a mine. Roughly, with us, whatever goes downwards is a mine, and upwards is a quarry, but there are great complications, for sometimes stone (for instance bath-stone and slate) is quarried below the surface, in mines, and sometimes minerals which should be mined, are quarried, on the surface ; so that most European nations define by the mineral, whether it is found above or below, but we can mine *or* quarry the same mineral, according to its position.

The phraseology of the expert miner is incomprehensible to any but the initiated. A ' bed ' is a layer of rock or earth, which is fairly continuous. A ' bed ' is ' thick ' or ' thin ', according to its height, from its roof (which is nearest to the surface of the earth) to its floor (which is nearest to the centre of the earth). The ' strike ' of a bed is its direction (technically, a horizontal line drawn on the plane of stratification). Its ' dip ' is its tilt downwards from the horizontal. The ' lode stuff ' or ' matrix ' is the stuff that holds the mineral that is being mined. (In the Rand, this matrix is a bed of conglomerate rock, and is called ' banket ', because that is the Dutch name for a sort of nut rock toffee—which it closely resembles.)

The country people in England are equally apt in bestowing nicknames on their own particular mines and quarries. The ' face ' in quarry or mine is, of course, where they are working, and an ' old face ' is where they have worked. The rock surrounding a lode or bed, is called the ' country ', and a large lump of this, isolated like a huge plum in a giant's pudding, and found in the matrix, is sometimes called a ' horse '. Slate quarriers have odd names for the small kernels found in slate.

' Vein ' or ' lode ' are words so hotly disputed by authorities, that the ordinary reader may be forgiven for sometimes misusing them. A ' vein ' or ' lode ' is a ' filled in ' crack ; how it is filled in, whether injected or permeated, may be controversial, but anyone who has seen fruit juice veining the cracked stoneware of a pudding basin, has seen a very good illustration of the result and at about the right proportion for some small veins. When a vein or lode is thick, and comes out in a long steeply sloping tunnel, it is called a ' shoot ', and the space where this is cleared out is called a ' pike ', or sometimes, loosely, a ' chimney '.

When a country miner speaks of ' working through a thin strike to a rich strike ', he probably means that the line he is following zigzags a little, and gets regularly thick and thin, much as you could imagine cement would, if it had been poured down a staircase—you would have thick places on the treads, alternating with thin down the uprights (hardly worth getting out, except that they lead to the thickening on the next step). A ' gallery ' is an endless passage-like working, though that again may be a misleading term in some pits.

Of old, they did not make careful scale maps of mines such as those of to-day, which are often sectional con- structions of layers of glass, that show all the various deeps, levels, shafts, and all ventilation and electrical systems, in fact, the entire mine at one glance. The old miner simply carried the geography of his smaller pit in his head. Not as a flat map, but in three dimensions, even as the shepherd of the hills carries the *form* of his mountain grazing, in heights an acreage.

It was once my privilege to be taken to explore an old mine in Ireland. The place was on a hill-side and the natural system of running underground streams and rising air had kept the air pure. Falls and caving-in had made some of the galleries impassable, and all records of the pit had been destroyed.

The expert who was surveying the mine agreed that it was one of the most unsystematized warrens he had ever penetrated ; yet, in that one day, an old Cornish workman who had crawled with us (carrying neither line nor level, and seemingly interested only in the perished timber props) had made a mental map of the pit, and was able to predict where many of the blocked galleries would lead and connect, and give estimates for levels that proved almost uncannily accurate.

The old method of getting the mineral out of the matrix usually involved breaking it by hand, crushing it with heavy mauls, and afterwards washing in running water. It is interesting that most of the various mechanical stamping machines that are now used all over the world have their prototype in the simple early Cornish stamping mill.

I have drawn you a very simple diagram of this early type of stamp, as run by the water power of some mountain stream ; you will see the same principle, though tremendously complicated and involved, behind the rows of thundering stamps that shake the ground in the largest gold mines in the world, and you will see it, also, behind the three or four little stamps which are beating away hopefully at the entrance to so many little gold mines in England.

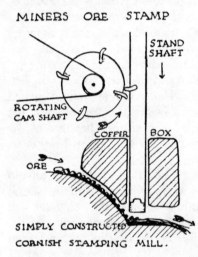

MINERS ORE STAMP

STAND SHAFT

ROTATING CAM SHAFT

COFFER BOX

ORE

SIMPLY CONSTRUCTED CORNISH STAMPING MILL.

Other methods of stamping and washing ore are interesting for the traditional patterns they have left behind.

Though these old mines were so small compared with those of present day, they gave much more local employment to other country workers. Timber had to be cut for their terrible system of moving ladders—a curious kind of double lift, like a tube of hollow boxes, cut in two parts, and as each part alternately rose and fell, the workman came up or down, by stepping alternately from one half-box to the other.

The wooden windlasses were made by special workers ; the panniers for man and beast that carried the ore were special panniers, even the ingots were specially made and specially stamped and weighed for loading from the mine to the exchange market. Great ventilating fans, like windmill sails, had the cloth specially woven for them. The heavy wooden suction pumps used whole tree trunks, hollowed out by hand boring, and charred with fire to harden and preserve them.

The traditional skill lingers in the people even as the traces of their work linger on the hill-side. The old tramlines on the hills can sometimes be tracked, long after the rails have gone, by the change in the sparse vegetation over the rotted wood sleepers; they cross-stripe the track for miles, to prove it is no ordinary mountain path, but will lead to an old pit or quarry. The old mauls that used to be made for hand stamping can sometimes be found in derelict worksheds: women and boys used them to beat down the ore into pans in the running water of some mountain stream. The little wagons of the miner and quarryman, and the iron rails and wooden 'tally-board' (see p. 126) are curiously unchanged among the larger alterations.

Smelting ore ceased to be a country trade when charcoal ceased to be used. With the change from wood to coal, smelting left the woods and became a coalfield industry. Now-

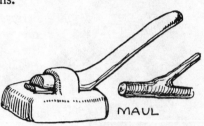

MAUL

adays you can come down off the moorlands that surround the Black Country and see the furnaces running out, and, glowing in the darkness, the great white-hot 'sow', and the little red 'pigs'. (He was a countryman in country England who once gave them names like these.)

For lack of space, I have had to omit any description of the work underground, and of the many small country industries that can be dependent on it, such as preparation of the timber, and utilization of the 'adit' or drainage water.

§ 2. *The Pitman*

The country people say that all metals 'pass into the blood', so that the inheritor becomes as sensitive to his 'family' metal, as a diviner's twig is to water. 'The mineral does soak into the blood of them that have to do with it, and being heavy in their water, gathers stronger father to child.' And this, said to me by a country miner, is believed in substance by hundreds of country people.

But mining is no longer a country trade, and the track ways of the old pits, the unsafe shaft-heads (dangerously gaping in the open moors), the old tools and implements that the miner once made for his own use, are all you will find left in the country places.

For, to-day, the miner has ceased to be a countryman. The out-of-work colliers, now massed together by the huge amalgamated colliery companies, netted in the electric tramways of the depressed areas, might well envy the sturdy countryman of old, who walked to his work in some small country pit or rode with his mates shouting lustily in lumbering stone wagons along their rough mountain track rails in the country.

BILLY COCK
AND CANDLE

That miner was an independent fellow. His bowler hat had a lump of clay to hold the tallow dip; his pick, clay pipe, and clogs were emblems of his prowess. He bathed in a tub of steaming soapy rain-water before his roaring kitchen fire and his woman sluiced the swill water, cold, over his head, as he got out. He was thick of skull, and strong of body, kind of heart, and he had an almighty thirst. . . . But, he used his own brain and it was good, though rough, and he played his own games and they were rough too, but good. It was his boast that his wits were as quick as his fists, and he kept both busy. The pet mice, the cage birds and pigeons that he loved were more than interesting toys. They were small, wise counsellors, and he valued their sensitive advice about conditions underground. Many a bird, finger-tame, travelled down the mine inside his greasy bowler hat (and they use birds, and mice also, for the same reasons to-day).

If, above ground, the agriculturalists chaffed the miner for ' tillin' t'aters with a pick ', at least his vegetables were a notable success, and he paid (for those days) comparatively big prices for his seed and for his special fancy in flowers. He was ahead of some farmers in appreciating breeding in vegetables, and his influence is another track by which we note his passing.

Pits were (and are) as individual as mountains. The workers would know them as good or bad, rich or poor, and they would swap news from one pit to another, news and warnings. The personnel of a pit in those 'smaller' days were known as individuals. Now, they are only calculated in numbers.

But though there is much of value left in the country-side, bequeathed by the country miner when he became a townsman, the present day miner has a sturdy inheritance from his country forebears. No miner has ever abandoned a fellow-miner, and every rescue gang has a waiting list of volunteers.

§ 3. *The Rockman*

Under this heading we include both miners and quarrymen. In the days before Government Safety Regulations, these men, owing to the dangers of their calling, had perhaps the strongest sense of personal responsi-bility of any workers. The majority of workings would then be small, each employ-ing a small number of men, who would be dependent on each other both for safety and output.

AT A ROCKMAN'S DOORSTEP

Under these conditions they became apart from other workers in the severity of their laws, which were made and enforced by the rockmen themselves, to safeguard their rights and their safety. Gathering naturally into certain districts where their work lay, such rockmen formed large fraternal communities, which soon set up their own courts and laws, and these were recognized to a greater extent than in any other trade. There remains to-day this special code of mining and quarry legislation.

It was obvious that in the dark underground, where such extremely technical work was concerned, even an expert could never explain the circumstances of a crime to outsiders ; unless the entire court went down the mine, they could not be shown the evidence. Only quarrymen or miners themselves could possess sufficient technical knowledge to judge the crime and it was for them to assess the punishment. Thus, for a miner caught working on another man's seam, the penalty was to have his hand nailed to his windlass. The wrongful stealing of the coal may have been proved, but if his fellow-workers knew him as a decent workman, perhaps newly come to the spot, or if some tilt or change in the line of the seam appeared to have misled him, then the nail would be no more than thrust between the skin of his fingers, leaving a slight scar to remind him to be more careful. But if he was known to be a dishonest workman, robbing with intent, that hand would be crushed past usage for many months to come.

That is a simple example of a frequently very complicated situation, involving perhaps such divided issues as mineral finding, when the finder, the owner of the land, and the representative of the Crown all had to meet and gain egress to the nearest highway by ceremonial usage, walking shoulder to shoulder.

These ancient mine and quarry laws, so intensely interesting and vital, are rather outside the province of this book, but they are mentioned because they were evolved not by industrialized robots, but by some of the cleverest brains inhabiting sturdy country bodies. ' The Rockman's Arms ' stands for more than a name on a country pub. The rockman character *was* rock, in its strength and solidity. Rockman and miner were a great force in the community of country England.

The true-born rockman (for they are born, not made) has always been one of the finest characters in England, with a farmer's patience, a woodman's imagination, and the constructive vision and balanced mind of a mathematician. Of old, without infringing the boundaries of his legitimate craft and often unable to read or write, the rockman could do wonderful things. Even to-day they do not easily put pen to paper, and probably there are few people more inarticulate, few people whose mental processes are less formulated. They have always used an instinct as completely unconscious as that of an Eskimo at a seal hole. Looking at a rock and its position, they will arrive at an equation demanding mathematical formulae far beyond their conscious calculation. They will say, putting a finger on the spot, 'The shot will shift it here,' but remain quite incapable of telling you how they arrive at that perfectly accurate judgment.

In proof, consider many of the remote old stone bridges of England, the small but well-engineered breakwaters, the old culverts and dykes. Confronted with such problems now, we should consider much mathematical calculation and heavy engineering apparatus necessary. Yet the old rockman, away out in the country, had no apparatus but what he made, and was himself capable of barely the simplest calculations. No two cases were ever quite alike, he could consult no formula, he. could only look at his rock, look at his problem, and *think*. . . . He *thought* very simply, till he ' saw ' how he could do it, then he acted very strongly; and the job got done.

Rock enforces thought before action. Rock is a very powerful master. I believe much of a rockman's calculation, even now, is of that same curious ' block ' method as old shepherds' countings; not a mental appreciation of separate numerals, but a mass appreciation, the same adaptation of numbers into units of time, space, and body that makes some footsore nomad describe the ' distance ' from his camp as ' four days '. A rockman's calculation of weight is more a sense of structural balanced equivalence than of numerical figures, more physical than mental in its origin.

Perhaps this ' thinking in block and weight ' simplified the rockman's vision, for study of many small buildings and structural engineering feats about the country will show most complicated difficulties, tackled and solved by his steady structural skill without line or figure on paper.

In old cathedrals and larger buildings, even the lack of known architectural plans does not blind us to the belief that there must have been such plans in the brain, if not on the parchment. We have found fourteenth-century drawings of workmen squaring and even roughly carving the stone near the quarry, to lighten its transport to the position it must fit in the finished work perhaps miles away, which argues pretty accurate fore-knowledge, and the regular master masons who travelled round from building to building, must therefore have been able to count on intelligent labourers.

Sometimes these large buildings show changes of plans and even altered or unfinished work. These great enterprises, the medieval equivalent of our largest engineering feats to-day, were doubtless directed by master minds and served by specialists. But in the country, the enormous number of satisfactorily completed lesser works is probably due less to the ability to transmit ideas by writing, or to the definite working out of plans by a superior mind, than to the country rockman's own mental equipment. They were strong in body and brain. In them was developed the completely English physical mentality.

Rock work made self-reliant men. Realize the isolated position of some of these workers. If unexpected difficulties occurred, there was no one to consult ; *they were country people working alone.* Their problems may seem easy to us, with our mechanical devices, and unlimited electric, or steam power, but to them, they were problems requiring thought and strength. More problems or harder problems required more thought and more strength : so they continued to develop both, fully. Whatever political and economic developments must take place in England, whatever facilities may now be available to transmit knowledge ready-made, let us not stamp out this ability in our people to think for themselves.

§ 4. *Quarry Work and Tools*

The distinction between mining and quarrying in England differs from that abroad, where as we have already said a *metal* will be mined, whether it occurs above or below ground, and a *stone* quarried, whether it is cleared from the outcrop or dug out from below the surface. In England, however, as already explained, we quarry above ground, and mine below ; so that the same stone or metal may be both mined and quarried according to where it occurs. Thus it happens that a miner's tools for a material probably resemble the quarryman's for that same material, only adapted for use in a more confined space.

Practically all modern excavating and boring is done by the compressed air drill, but this drill is only a mechanical form of the old manual jumper or fritting iron (see drawing on p. 118), which is still in use for its own particular work. A jumper is frequently used where it is impossible, or inconvenient, to ' lead ' the mechanical drill. Recently, in testing for reopening an old neighbouring quarry, jumpers were used to avoid the expense of bringing up the engine and the drill, yet the time taken was very reasonable. In country work the mechanical tool is often only an adaptation of the hand tool which antedated it and both may be used together.

The blasting charges change, but not the simple wood brush which clears the hole before putting in the charge. That is still a bit of stick hammered out into a rough brush, whether the hole is made by a modern up-to-date rock drill or by a fritter. The wedge, the sledge-hammer, the pick and spade, and all the strong simplicity of the rockman's tools continue in use in local quarries. It is impossible to give a full list, but these are representative, and if you understand their uses, you will be able to follow more complicated workings.

Jumpers are of as many varied shapes, weights, and lengths as the rocks to which they are adapted. A small hole is first bored with cold chisels (called by some rockmen ' a fritter '). As soon as the hole is deep enough, the work is taken over by a jumper, which is really only a long chisel, and which jumps round in the hole, boring downwards at each blow. Some jumpers are weighted, swelling out to a heavy well-balanced thickening in the centre. Sometimes a second man holds the jumper steady during the early stages of the work. The pointed jumper in the diagram (this was used for slate) has a wood and leather holder slipped over it for this purpose. If the hole is being driven horizontally, jumpers with the

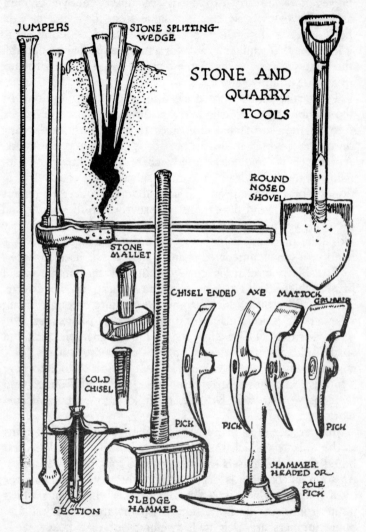

JUMPERS STONE SPLITTING WEDGE

STONE AND QUARRY TOOLS

ROUND NOSED SHOVEL

STONE MALLET

CHISEL ENDED AXE MATTOCK GRUBBER

COLD CHISEL

PICK PICK PICK

SECTION SLEDGE HAMMER HAMMER HEADED OR POLE PICK

weight set at different distances down the shaft are used in turn to get the right balance as the work progresses.

As the jumper bores downwards, the hole is usually filled

with water, which assists in keeping the temper of the chisel and carries the rock particles in suspension. Therefore a leather flap is sometimes slipped over the jumper (see diagram) to keep the water in the boring. Holes up to 14 feet deep may be bored by suitable jumpers. When it is necessary to clear out the hole, a chip of wood hammered out into a rough brush with a dab of tar or grease on the end lifts out the stone chips. A jumper, with a chisel end for boring, should not be confused with the ' bar ' used for shifting.

Once the holes are the required depth, they are used either for a charge of explosive or for wedging work. The wedge may be of wood or metal according to the stone and the position. Dry wood wedges driven in and a bucket of water to swell them are sufficient to remove very large pieces of rock. But more usual are the metal wedges. They are driven in line, and the worker goes along the line backwards and forwards, ringing each with a driving blow of his sledge. The note goes higher and higher, taut as a bell, till suddenly the note drops, dead: it is done, the block is separated.

I know no more satisfactory sound in country work than the noise made by a good piece of wedge-driving in a quarry. It is the mounting tension of endeavour made audible, followed by the quiet finality of achievement.

In the sketch is shown a very complicated wedge known as the ' feather wedge '. It has various forms and uses. The feathers may be blocks thick enough to arch across a wide crack, or as fine as the sliver cut from some old shovel and inserted to balance a wedge.

Having detached the rock by jumper, drill, blast, or wedge, the debris now calls for other tools, the commonest being the rockman's pick. These are issued in sets to each worker. The quarry or mine smith sets and often practically makes these, and tempers all the men's tools according to their special requirements and the nature of the rock where they are to be used. He is one of the most important persons in the quarry (see p. 118).

In the diagram are shown four different picks, or rock tools. They only represent a general type for each make of pick, as there are several hundred different picks of each type; probably no two quarries in England utilize exactly the same pick, and after the general type has been adopted, the quarry smith usually adapts it to suit the workers' preference and usage.

Technically, No. 1 is a chisel-ended pick; No. 2, similar but with the chisel end set parallel to the shaft, is a pick-axe; the last two are variously named shovel picks or grabbers. (A lighter variant of these served out to roadmen to be used like hoes are called ' grubbers '). Some are (erroneously) called mattocks. Upright No. 5 is a hammer-pick. The small, exceptionally strong-handled shovel in the top right-hand corner, with a pointed end for inserting under the stones, has a slant set handle, the angle of which is exactly worked out between the horizontal surface of the ground, the height to the shoveller's shoulder, and the weight to be lifted. The subtle and exact ' pitch ' of this handle makes hours of difference to a man's work during the day.

Other rockmen carry on the skilled and interesting country jobs of boring and sinking. These jobs vary from plain well digging in chalk (where the apparatus used is a stick and string, to scratch out the circle, and later two buckets, one spade and a coil of rope with which the men ' just dig themselves down '), to mighty engineering enterprises, in which huge drills bore down through the different rocky strata with changeable cutters. It is now possible to do this at varying angles, making it feasible to drain oil from directly inaccessible places. The cylindrical cores brought up in the bore tube are perfect geological sections and sometimes the boring is done only to collect this evidence.

An example of pressure boring are the artesian wells in the chalk bent of London's country, which have revolutionized the watercress industry and reconstructed many market garden areas. A worker connected with oil and water pipes is the plumber who still uses a countryman's ' moleskin '.

§ 5. *A Limestone Quarry*

We will take quicklime as an example of things made from stone, since that enables us to describe one almost diagrammatically simple quarry.

This quarry is situated upon a hill-side. The limestone is of good quality for burning and has been worked for at least 100 years, so that gradually the work has eaten away half the hill, and on one side, where the limestone workings adjoin an old derelict copper quarry, the boundary roadway is left 100 feet in the air on a knife edge of rock. It would be utterly impossible ever to widen *that* road. (This point is interesting, as the same problem crops up regularly in quarry districts.) Near Carlow in Ireland an inversion of the process quarried out, for the stone, a slit of a road, deep as a trough, through

THE MOUNTAIN ROAD

the solid rock. For 10 miles the rock walls rose sheer, 10 or 20 feet high in some places, barely 10 feet wide, so that carters had to yell loudly before entering the cut, or risk having to 'back out'. Another ridge of road, this time of chalk, occurs near Maidstone, in Kent.

The limestone in the above quarry slopes back up the strata of the hill. The face is stepped backwards for working as there must be no dangerous overhang. The ground on the top of the quarry edge is also cleared back for a statutory number of yards, so that no sudden shift of earth or wind-loosened tree might come sliding down over the edge on to the workers below. At intervals along the base of the face are levelled platforms of rock debris on which the ends of the rail roads run out, and allow the empty trucks to wait level while they are loaded, with the safety point before the slope pulled up to prevent them running away.

The old platforms and the old face where they have finished working are overgrown with small mountain shrubs, and myriads of flowers, wild rose, and honeysuckle, and ragweed, patches of purple comfret (perhaps from the old quarry horse stables—it was used as a horse medicine), big white bladder campion, and in the autumn the finest blackberries among the gold and red, for the great stone cliffs make shelter from the east wind and hold the warm sunshine as in a big white basin. Sheltered quarry flowers are among the finest and earliest, and quarry blackberries ripen hot in the sun as in a greenhouse. Birds nest there, regardless of the blasting, and some of the men take the straight brier rose stocks, and bud on to them, making standard rose trees for their garden. These standards always do well, for the root stock is so well acclimatized.

Wandering through this rocky, flowery 'waste' are small trackways that lead, apparently, to the blank face of the rock. But they are not dead ends; each track leads to a quarryman's short cut out of the quarry, and often these 'accidental' little track ways made by the men have later become regular footpaths; many times they are the foundation of future village roads. The placing of our hill-side villages among quarries would be a real problem for town planning.

To return to the working part of the quarry; the trucks stand at the top of a long, sloping tramway line that goes down, and around, and down, to the roadway and kilns below. The distance is about half a mile, but it does not cost anything to convey the limestone that distance, nor is any hauling machinery used.

The diagrams show the usual arrangement of gravity incline or counterpoise that is used in all quarry work. You will see that it is exactly similar to the old windlass well, with its two buckets (into which, if you remember, the wily Brer Rabbit induced Brer Fox to clamber and by his heavier

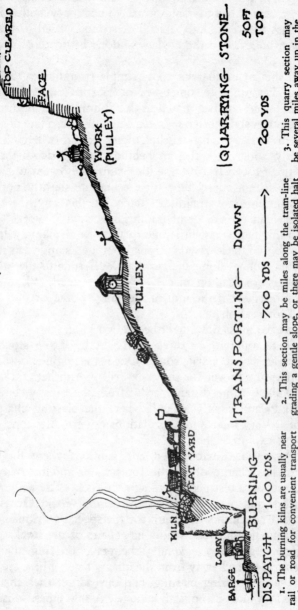

DIAGRAM OF WORKING LIME QUARRY

TOP CLEARED

FACE

WORK
(PULLEY)

PULLEY

FLAT YARD

KILN

LORRY

BARGE

|QUARRYING STONE|

50 FT
TOP

200 YDS

3. This quarry section may
be several miles away up in the
mountain.

|TRANSPORTING DOWN|

← 700 YDS →

2. This section may be miles along the tram-line
grading a gentle slope, or there may be isolated halt
sections where a horse is used to pull across a level
crossing (roadway).

|BURNING|

100 YDS.

1. The burning kilns are usually near
rail or road for convenient transport
and constitute a separate industry,
complete in itself.

DISPATCH

weight haul Brer Rabbit up !) The loaded stone wagon is heavy, and it has to come down. The empty wagon is light and has to go up and be refilled, so one is tied to the other over a pulley, and the haulage is done by force of gravity.

In this small quarry, it is a simple straight run (as shown in the diagram), but sometimes for quarries remote in the hills, several miles are negotiated in this simple way, and it puzzles and amuses strangers to see, on a bare hill-side, a line of sensible little trucks apparently rattling along from nowhere. There is usually a man standing on behind, holding down the brake with one foot. If, on these long runs, they want to slow the pace before a curve, they tilt the line up slightly, and make the running on a straighter stretch. In this way, the country quarryman does a lot of unofficial surveying work.

In this particular limestone quarry, as in many, a horse is kept to pull the lowered wagons from the ' station ' at the foot of the incline along the last few level rods to the kilns and crushers (or loading place). Here also road, rail, and canal transport meet at the foot of the hill. Some of the stone is used for road metal.

But we must now describe the working.

There are two or three sheds, built of the stone, with corrugated iron or slate roofs. One is the engine house, which supplies the power for the rock drills. Another is a separate store shed, some distance away from the quarry, built very strongly and closely locked, where the blasting charges are kept. Very interesting regulations govern the construction of this explosive shed.

There is also a tool-shed and smithy, where the quarry smith has charge of all the tools used, and in many cases designs the tools specially to suit the rock. The *temper* of the simplest quarry tool has to be adapted to its rock. Occasionally the quarry-smith will design for the special workman, and at any time he undertakes running repairs of the stock.

There are also several ' shelters ', or dug-outs, their openings turned away from the quarry face. These are shelter for the men during blasting, and one, larger than the rest, is fitted as a mess-room with a stove. A few pipes, loosely laid on the hill-side, carry off the surface drainage, and some deeper permanent springs are piped to cisterns near the engine house.

Let us now watch the work.

The rock drill has been whirling away for an hour or so. It works somewhat on the principle of the jumper, but often a bore tool takes out a clean section or core. The boring varies with the rock; an average $1\frac{1}{2}$ in. diameter is in use in this quarry. The *depth* varies with the depth the charge is to be planted. Dynamite explodes downwards and the charge may be required either to split off a rock or shatter it completely.

The silence after the cessation of the drill is almost tangible. The vibration and echo within the circular quarry has so long diffused the sound that it has seemed to buzz from all sides. Now, like the quivering of water in a bowl, the air subsides, and the little sharp sounds of iron pick on stone, or the voices of the men, sound isolated and musical. In the quiet, Bill, the charge-man,

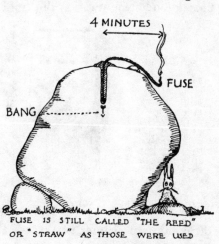

4 MINUTES

BANG

FUSE

FUSE IS STILL CALLED "THE REED" OR "STRAW" AS THOSE WERE USED

is going around fixing the blasts into the holes. The law forbids him to use any instrument of steel or iron lest it should strike a spark : he uses copper and wood. The fuses, like little bits of black string with silly frayed-out ends, look inadequate and small, like elephant tails dropping from the huge rocks. As soon as Bill is ready, a bugle blows, and instantly everyone in the quarry makes a bee-line for the nearest shelter. This is the first law in a quarry. When the bugle goes—take cover. And within half a minute, the stone dug-outs are full of men, and the quarry deserted, but for Bill, the charge-hand, who lights the fuses and is last to leave.

Boom, boom, boom . . . three, four, five, Bill is counting

the number of blasts that he has put in. Each explosion shows a puff of white dust; some, against the side of the rock, are followed by an avalanche that brings down more of the cliff. Many are on the ground at the base of the face, splitting rocks which have not broken in falling and so are still too large for the men to cope with. Sometimes Bill puts in a charge to change the position of some inaccessible or dangerously balanced rock. . . . When the echo of the last boom has died away there is a pause, one minute, two minutes, three minutes, and then the 'clear' bugle sounds, loud across the quarry, 'all clear,' and the men in the shelters rise off their heels and unbend and come out. . . . But until that bugle is called not

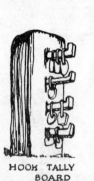 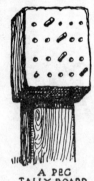

HOOK TALLY BOARD   A PEG TALLY BOARD

a man will stir. Along the face where he set the fuses goes Bill, checking up each fuse that he put in, learning by results and experience, storing up the knowledge of charges and the ways of rock that make him the efficient workman he is.

The period in shelter before the 'all clear' and many other details in checking and double checking the number of charges issued, are all statutory, designed to prevent a possibility of a charge remaining unexploded in the rock. The face *must be clear of explosive* before work is resumed. When the men go back, those who were only 'loading' continue where they left off, but others have to study what the blast has done for them, and how to tackle the resultant fall. The nature of his work keeps the rockman constantly confronted with fresh problems, never twice exactly the same.

The men work in gangs of four or five. They load their waiting trucks very methodically, fitting the rock like a small stone wall on the open side of the truck and piling in and up towards the centre. Very neatly balanced loads, and carefully tallied for as they are sent down, each truck as it rattles away carries its own tally hung on a hook behind: the man at

126

the bottom takes the tally off the truck and hangs it on a peg, and notes that a particular gang sent down another load to its credit; and the rail-man braking the truck as it passes down the run moves a wooden peg one further along a row of holes in a wooden board, indicating the passing of another load. Paper and pencil are pernickety things, and the last tally system in England will still be working in the last English stone quarry.

I found the derelict tally board of an old quarry in the Arenig mountains. The old carved pegs had grown in tight, holding immovable for ever to their testimony that 'Hughes and Henry Roberts had done 2 and 3', and 'William Roberts 3', but 'Henry Williams had done 5.' It was over a century old, and now, Hughes, Roberts, and Williams and their sons like good Welshmen rest in the little chapel graveyard below, but while the night wind burrs over the brown winter heather, and when the cold silver rain of spring brings back the sheep, always, all alone on the wet mountains, to the wheeling curlews and the trembling rainbows, that old tally board still staunchly testifies that Hughes did 2, and Roberts did 3, but Henry Williams, *he did 5*.

FINIS

The lime kilns of the quarry where the limestone is burnt are very large and usually three or four of them are in action. Formerly, before motor transport was available, there were many more small kilns working up and down the land. Every district that could produce any sort of lime on the spot for its fields did so. There are thousands of caved-in old lime-kilns making humps about our farmlands.

To burn limestone the rock is packed into the kilns over a firing of wood and coal, lit from below. The flues can be regulated for draught. The kilns burn with a transparent blue waving flame by daylight, and at night they glow a round pool of fierce red heat, and the thick yellowish smoke that pours out is lit from below a deep soft red, like blanket banners, billowing away in round clouds. A road runs below the towering kilns, and they light the small mountain valley for miles.

The rock-like rampart of stone wall enclosing the kilns is permeated with a low steady heat, and house-martins build their nests in the wall; I should think it must incubate the eggs. . . .

§ 6.   *A Blue Lias Quarry*

As an example of the simplest rock work, carried out without any ' power ' by a few workmen, I have chosen a blue lias quarry from the West of England. Of course, blue lias is also worked on a big scale, but its nature is what the men call ' tractable ', that is accessible and easy of manipulation. In this case a small outcrop occurs very close to a main road, and the men work the stone with the minimum of apparatus and simply transport it to the roadway on a strong wheelbarrow.

I want to break down the idea that a country job is necessarily less efficient just because it is carried on in a small way. Many country jobs depend on the relation between the material and worker. No amount of ' plant ' and machinery would alter the result, it is merely a question of the size of the job and the apparatus required.

Efficiency is not the monopoly of big business. Often it is expedient to work some small quarry such as an outcrop of nicely slabbed sandstone for a wall, or an outcrop of rock broken off and reduced to road metal for local use, on a really small scale ; in no case think the less of the rockman's skill because he is working in a small quarry.

The name ' Blue Lias ' is an example of the countryman's nomenclature. In books you may find it described as ' the equivalent to Hettangian, Sinemurian, and Pliensbachian, or Charmouthian (in the restricted sense, after deducting Domerian) '. But the workman remained unperturbed by all that. He said, 'It's these *blue layers* they mean,' and 'blue lias ' it is.

On a large scale, blue lias is one of the finest building stones we have, and is much used for paving work. As a form of limestone, it is sometimes worked for cement, and in some

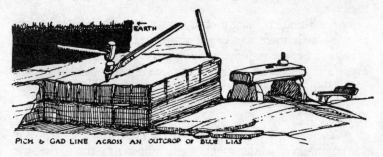

PICK & GAD LINE ACROSS AN OUTCROP OF BLUE LIAS

districts is in demand for bakehouses. It has a clean, level finish, and makes a good bake-stone or hearth. The lias being a sedimentary rock, the cleavage is convenient, as in slate.

In the quarry I describe, a man with a shovel clears away a foot or so of surface earth and turf, and they get down to the

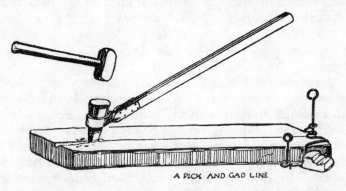

A PICK AND GAD LINE

first layer with very little fuss. Having cleared this earth back, as wide as the slab they want, they get a poll pick and ' nick ' across in a series of small holes, which will cause the lias to snap off along that line.

Sometimes this is done with a ' pick and gad '. The simplest holds for the gads are two light wooden handles and a twist of leather. With this the gad is drawn by one man, while another follows along hitting it on the head—a quick and simple process. For a deep hole they use a 'jumper'; a jumper is a very good name for it, for, as has been explained, it is a very long chisel that jumps round and round in the hole.

For splitting off large pieces, if a small charge of explosives is not used, there are other simple methods. Sometimes a bucket of water judiciously poured will be sufficient to expand and crack off an undercut piece of rock. Another trick is to drive in a line of old wood blocks, wet them (or a shower of rain may do it for you), and the swelling will force a fracture.

As the stones from this quarry were being used to complete the paving kerb in the street of a neighbouring town, the size and finish were known, and the slabs could be completed ready for setting in the quarry. For this, they were built up on wood blocks, conveniently high, using a long iron bar as lever, the size being marked off with a steel rule. They were cleaned down to a finish with a cold chisel and mallet. It only remained to trundle them a few yards to the roadway to await the collecting lorry (Illus. 29, 30).

§ 7. *Dry Stone Dykes* [1]

Many great stone dykes or walls still stand firm for hundreds of miles across the moors. In places they climb down hill-sides so steep that to-day we should consider them impossible to build without cranes and engine power. Certainly the dykes could now be built more quickly using mechanical power, but I stress the point that these great works were carried out unaided by ordinary country workmen, because they are an excellent example of how such men have used the principles of later mechanical ingenuity. Dyke work was intensely skilful, and demanded great judgment and foresight. The instinct for rock, the sheer animal strength and sense of balance which makes mountaineering a joy,

[1] They are called "*Dry* stone dykes" or "*Dry* walling" as they are built without any mortar.

was part of the everyday job of the quarry and dyke men. Truly, in these stone dyke workers the English engineer has a noble ancestry.

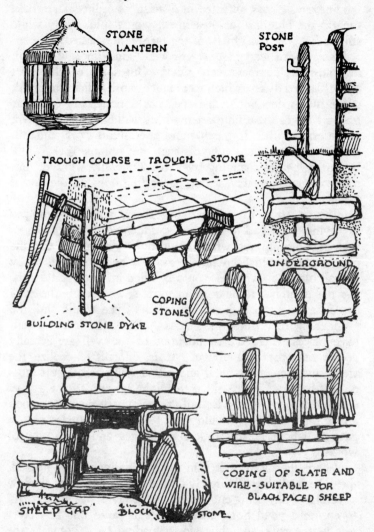

STONE LANTERN

STONE POST

TROUGH COURSE — TROUGH — STONE

UNDERGROUND

BUILDING STONE DYKE

COPING STONES

SHEEP GAP

BLOCK STONE

COPING OF SLATE AND WIRE — SUITABLE FOR BLACK FACED SHEEP

The work was undertaken in summer, for in winter the wet rock was too cutting to the hands and snow made the

131

work impossible. The material was always the local outcrop. The making of any dyke, therefore, varies as much as the rock itself does. The rock which was most easily worked and made the finest walls was obtained in districts of sedimentary rock, such as the blue lias just described, which splits easily while still ' raw ' (i.e. straight from the face, not weathered).

In slate districts these layers were sometimes very thin, in sandstone, comparatively thick. But some of the most well-balanced dykes, which were much more difficult to build, occur in districts where the rock has a hopelessly irregular fracture. Even great flints are used for wall building in certain localities, but flint is much more specialized work and not nearly so extensive. The distance that suitable rock would be carted, before changing over to a less easily worked, but more available quarry, would depend entirely on the lay of the land.

When a track for a dyke was fixed, much preliminary surveying work had to be done. Sudden small irregularities, such as a jutting crag which could not be built into the line of the wall, would be removed by bore and wedging, or by a small blast. Soft places would be filled in with rock debris, and this often had to be done an entire season before the dyke could be built, since the weight of the dyke would sink the foundation rock into the moss. Some boundary walls have places where the foundations of the wall are actually deeper than the wall above. It is difficult to realize the work involved in draining and pile-driving across soft moss or peat hagg. Five or six men might be needed for a very difficult passage, otherwise only two country men undertook a task for which we should now send motor lorries with a gang of labourers, mechanical appliances and a director of works.

Where the moor was firm no foundations were needed, the dyke could be laid directly on the surface, only the shallow covering of turf being cut and removed. This was done partly to ensure a good bottom and partly because this turf layer was very fertile, and thus worth loading back home as the carts returned empty from bringing up the stones.

The track having been marked out, suitable rock was then

carted to the spot and laid down alongside the track. There was some skill in this spacing. Roughly they allowed a ton of stone to 1 square yard, varying of course with the variety of stone. The dykes averaged 5 feet high, tapering from 2 feet at the base to a coping stone slightly less than 2 feet on top. Since it was impossible for the builders to carry the stone any distance, a fairly accurate amount had to be laid down convenient for their use. If there was any doubt, the extra stone was deposited slightly uphill—it could then be easily rolled down.

The dyke men worked in pairs, one on either side of the wall. They had a frame set upright to hold the plumb line, and a cord to give them the track. A spade for levelling the turf and a hammer pick for the rock was their entire outfit.

Their skilled work consisted in lifting and setting the stones with a slight cant outwards (for drainage), and a strong binding through from side to side of the wall. The two faces of the wall are kept as even as possible.

Some workers (you will see their walls in the Dales and elsewhere) used to leave ' through stones ', as they called the pieces which reached from one side of the wall to the other and served to bind it. These ' through stones ' jutted out on both sides, so that the lines of them can be seen sweeping either side of the wall. The same principle is used when building the stone step-stiles so common in dyke districts.

An important fact, not usually realized in considering the construction of these dykes, is the end-to-end pressure. After a dyke has been finished to its complete height for a good length, the workman goes back along it, driving in extra wedged-shaped stones wherever he can, especially at the ends of level layers where he has sometimes left a small crevice to receive the wedge. Thus a slight end-to-end pressure is set up along the length of a dyke, in addition to the downward pressure of its weight, and the whole dyke becomes a structural building and not just a pile of stones.

On sheep farms all stone dykes have small culverts or bridged openings made at intervals to let the sheep through. These openings run about 3 feet square, though Welsh sheep can do with smaller passages, so the builder tries to keep

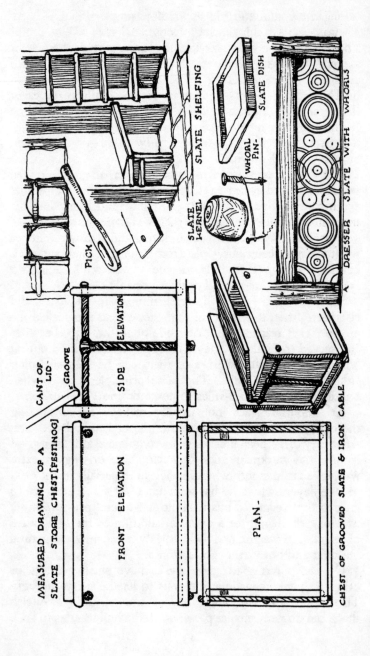

MEASURED DRAWING OF A
SLATE STORE CHEST [FESTINOG]

CANT OF
LID

GROOVE

FRONT ELEVATION

SIDE ELEVATION

PLAN

PICK

SLATE SHELFING

SLATE
KERNEL

WHORL
PIN

SLATE DISH

A DRESSER SLATE WITH WHORLS

CHEST OF GROOVED SLATE & IRON CABLE

a long stone handy for a lintel. Otherwise, he has to corbel out his wall, which is not so satisfactory. Frequently a single round stone is kept handy to stop the gap, or a thorn bush is pulled through, tail first. Above the coping stone, the stone dykes are finished according to the type of animal they are to pen.

The diagrams give a one-up, one-down, coping, very good against the average clambering goat-like jump of a mountain sheep. Not because the sheep cannot jump so high, but because there is no landing room between the upright stones. (Sheep jump up and down—they don't leap over.) Clever black-faced sheep, and some are very enterprising mountaineers, or ewes anxious to get back to their lambing pastures, make light work of even this strategic coping, so sometimes uprights of wood (or slate) are wedged between the coping stones, and a single wire is stretched along the top, just too high to clear and just too low to creep under (p. 131).

§ 8. *Slate*

There are numbers of countrymen employed in slate quarries. The large quarries such as the Bethesda in North Wales are well known, but few people realize that there are hundreds of small slate quarries working hidden in the mountains. In some cases the men ride in the slate trucks several miles up into the mountains to the face.

The work is very skilled and highly specialized. Slate knapping forms one of the regular competitions at any sports meeting in a slate district. Often slate competitions are run in conjunction with sheep dog trials, since both shepherd and quarryman work on the same mountain.

Slate handling is a matter of knack and skill rather than strength, and the facility for the finest work is not easily acquired. Some of the old workmen develop a wonderful sense of touch and balance and will split slate true to a hair with a very light blow exactly placed. The curious slender strength and balance of the slate workers' tools are shown also in the very characteristic and structural quality of all slate designs.

The slate chest or coffer (shown in the drawing) is typical of the slick structural sense in slate. The lid is so balanced that the extra 3 inches on the back side of the coffer supports it neatly in the two small grooves cut in the side pieces. The iron tie was a piece of quarry cable.

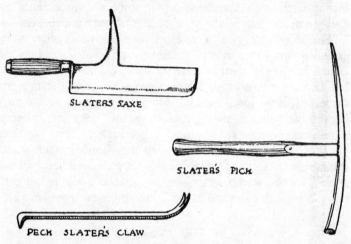

SLATERS SAXE

SLATER'S PICK

PECK SLATER'S CLAW

The slate dish was probably a milk cooler and cream pan, since slate is deservedly popular in dairies.

In a slate district large unworked slabs of slate are used structurally for building purposes such as the partitions in a stone house ; the shelves and sinks might be all of slate and even the flooring and steps.

§ 9. *Sanding and Hearthstoning*

White sand and grey sand !
Who'll buy my white sand ?
Who'll buy my grey sand ?

Old Round

As through the streets I takes my way
With my bag at my back so gay,
Crying out ' hearthstones ' all the day,
' Hearthstones ! ' a penny a lump
' Hearthstones ! ' and ' Flanders brick ! '
A penny a lump, a penny a lump !
Who'll buy, buy-y-y-y-y— ! ? "

136

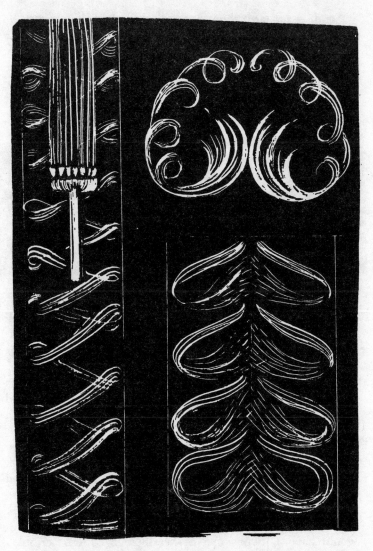

PATTERN OF SANDING FLOOR

WALKING BACK-
WARDS DOWN A
PASSAGE — ONE
HAND STREWING

[Above] TWO HANDS AND TWO
COLOURS FOR A DOORSTEP
[Below] A WIDE PASSAGE TWO
HANDS STREWING

Sand and hearthstone (both more used in the North than in the South), are now town industries. But they formerly employed a great many country workers, as all the floors of Northern bar-parlours and farmhouse kitchens were strewn with sand (whereas in the South they used sawdust) and Northern hearths and steps shone white from their daily hearth-stoning.

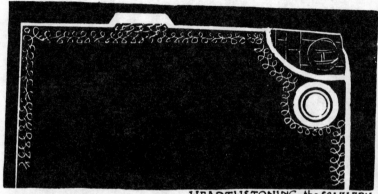

HEARTHSTONING the SCULLERY

The sand, red, yellow, or white, was strewn in patterns so intricate that it was said a clever barmaid could convey a message to her lover in a sand pattern. The sand patterns shown on the previous page were drawn for me by an old Yorkshire woman, who, incidentally, crushed her sand with an ancient maul from a neighbouring lead mine.

Hearthstone had a great sale in the North: it was often given in exchange for empty jars and bottles. Flanders brick was a yellow version of it. After the floor or step was washed the brick was rubbed over it and the whole smoothed white. Sculleries and passages which were only washed once a week were sometimes elaborately patterned. A wash-tub or churn would have a joyous mat of hearthstone pattern scribbled round it to show it would not be shifted out for another seven days.

CHAPTER FOUR : METAL

§ 1. *The Country Mechanic*

MECHANICS ARE a race apart. Scientists admit no national boundaries; neither do mechanics. I have never found a man of any nationality who 'had a liking for machinery' lonely or bored. If their inventive instinct was temporarily quiescent, they were still mournfully happy, grumbling their way to spiritual uplift with a small oil engine. Even the lowliest mechanic, if he wasn't taking his machine to pieces, was contentedly 'watching it work'. And all like-minded mechanics were brothers unto him.

In England, the average of mechanical ability is remarkably high. Foreigners often comment on the variety of technical apparatus, electricity and wireless 'parts', machine tools and engine 'spares' for sale in ordinary general stores, goods that would elsewhere only be known and purchased by the specialist workman.

In the country, the technical facilities are naturally small. The countryman, hitherto, has not had electricity in his house, no machinery in his field, and takes slowly to mechanics. It is often latent in the country boy, but he has not the means to satisfy his interest.

Nowadays many country engineers lament the difficulty in advising lads as to the best chance to get an all-round mechanics training. For the motor engineer, the country garage often offers wider experience than the town, where the work frequently degenerates into cross-indexing spare parts or doing one type of job only. So many a small country garage serves as jumping-off place for young men wanting to work out ideas, or inventing some gadget, for which they want to make working models on the quiet. In engineering, as in other branches, some of the best new ideas are made in England, in the country.

Why then with such good material in the country worker, do so many large works seem to turn out such feeble adults? A great engineering workshop is only a magnified edition of a small country workshop. Neither job nor workers are fundamentally changed, but they may too easily become commercialized out of recognition.

I was a factory hand myself for five years, one worker among eight thousand, gathered under one glass-roofed township. It was an engineering works, in size as far removed from our own little country workshop as anything could be. But though unlimited power was available at a switch, yet the lathe at which I worked differed in nothing but size from the small treadle lathe bolted to the stone floor in my father's workshop. The grindstone was a replica of the stone we used, save that the material was adapted for running at greater speed than our simple water-wheel. Even our little oil engine which coughed and worked so temperamentally seemed to have its equivalent in the huge power house (judging by the workers' exclamation, 'What are they doing at the bloody power house,' every time the output dropped).

The lubricant that overflowed the drip tray, the turnings that writhed upon the floor were familiar from the small shop, and even the cleaners, seemed in no wise different in effect from our small 'cleaner' boy who came at odd times for twopence; both were told 'not to touch anything'. In the huge factory, no peaceful hens wandered, pecking, through the open door to nest in the tool-box, but the factory cat came and had her kittens on my folded coat in just the same homely way. The forge was still the meeting-house on a cold morning, the smith a skilled master in extracting foreign bodies from your eye.

There were the same tea-brewing intervals (the married men brought milk in bottles and bachelors kept condensed tins in their lockers), the same incorrigible small boys who vanished on a three-minute job, returning, out of breath and full of virtue after twenty minutes. Large or small, the workshop characters were all the same. The old, old shop jokes, old as iron and tough as nuts : the cap carelessly dropped in the roadway, hiding an 8 lb. bolt, to bump the toe that tries a malicious kick—the compressed spring, coiled into the coat pocket—the magnetized feelers—all the same jokes, all the same jobs : the English mechanic everlastingly the same.

But . . . since the factory was larger than the workshop, the machines were larger, the power more powerful, and the achievements immeasurably greater than in our tiny country works. Then surely the opportunities for mental advancement should have been greater to match enlarged opportunities ? But no, the only thing in that huge enterprise which I found smaller were man's opportunities. Restricted at every turn, cramped and confined by rules, a worker could not even leave his lathe to re-grind his tool without coming up against the Management for leaving the lathe, or a Union for touching the grindstone.

Admittedly the War years were exceptional, and it was twenty years ago, but one still sees clearly that the lad from the small country works, an independent worker, ' small ' by reason of his lack of capital, his small apparatus, his small plant, and his small output, has nevertheless a wider training, greater opportunities, and a much larger outlook than a man restricted to one small process in a large factory. The shop is smaller, but the man greater.

The larger the town factory, the more impossible it becomes for any worker to get an all-round comprehensive training. It becomes increasingly difficult for even an intelligent workman to study, experiment, and invent freely as he did of old. Nowadays, for every young engineer who wins through to fully developed individuality, there must be hundreds, potentially as good, for whom it is increasingly difficult to study thoroughly even their own special branch.

§ 2. *A Country Smithy*

Under the spreading chestnut tree
    The village smithy stands,
The smith, a mighty man is he
    With large and sinewy hands.
And the muscles of his brawny arms
    Stand out like iron bands.

The brawny arms of that particular smith make us forget that a shoeing-smith takes an examination requiring anatomical knowledge as well as a very high average of technical ability.

Here are some specimen questions from a recent farrier's examination paper :

'State why the Corium is so vascular and give the names of blood vessels found in it.

'Describe how the horny sole flattens under weight and prove your answers by quotations, giving names of your authority.

'State what in your opinion causes a Kerotoma and how would you proceed to shoe the foot with this disease.

'Describe the growth of the wall, giving its chemical composition, its qualities, good and bad, and the position of the horn tubes in the wall, and give the ratio of heel to toe in a well-balanced foot.'

In addition the candidate must give practical demonstrations of very astute diagnosis, and carry out extremely skilful manual work. So that the urban estimate of a village blacksmith is entirely inadequate.

There are at least 100,000 skilled smiths working in England. The popular conception of a smith's work as coarse and heavy is justifiable, because much of an agricultural smith's work *is* heavy, and its durability more important than its appearance. But work is not clumsy if suited to its purpose, nor coarse wholly on account of its rough finish ; there is as much skill in shaping heavy intractable material to an exact hundredth of an inch as a softer, lighter material to a thousandth of an inch. The

A FLAT SHOE FOR PLOUGH-HORSE

SHOE WITH SLIGHT CALKIN

fineness of some smiths' work is almost incredible, considering that it is executed with tools in themselves heavy enough to withstand the usage of the forge. The technical skill and metallurgical knowledge of the average smith, in addition to his physical training, produces a very ' all round ' man. The casual observer, seeing only the crudeness of the forge and the tools, does not in the least realize the accuracy needed in their handling. In large factories it is an ' exhibition turn ' to bring the steam hammer down with the force of many tons' weight delicately adjusted so as to crack an egg. A similar trick could be done by many a blacksmith from equally delicate adjustment of muscular force and keenness of eye.

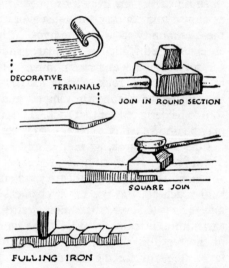

DECORATIVE TERMINALS

JOIN IN ROUND SECTION

SQUARE JOIN

FULLING IRON

On account of the quick red hot nature of the work at the forge, most measurements must be carried in the eye. To watch a trained smith turning out a line of exactly curved and matched iron palings is to see co-ordination of hand and eye that would be marvellous in a worker in fine glass ware. In iron, the exactitude of the balanced symmetry is often concealed in the sheer strength and weight of the material.

In early periods in England there used to be perhaps one smith to every handful of people, and he was probably the most indispensable member of the community. Among the earliest feudal records of workmen becoming independent contractors were those of the smith, who, by reason of the weight of iron, smelting facilities, or organizing ability, began very early to take on yearly contracts, even such simple contracts as ' doing all the shoeing for a certain establishment for a complete season'. I was interested to find the modern smith also likes to undertake the shoeing of the local horses by the year. So much of the health and strength of a horse depends on his adequate shoeing, and so often can a hurt, damaged, or misshaped foot be cured, or some defect in a horse's gait be alleviated by the studied work of the farrier, that to a great extent the condition of the horses in any district is a verdict on the skill and character of their local smith.

Of old, the craft of the armourer and other subdivisions of smith work were even more distinct than they are now, but as always the countryman must work to satisfy the needs of the community in which he lives. There are, of course, the specialist smiths, such as the quarry smith, who works entirely for some definite quarry or mine. Such a smith specializes in his particular rock, and is learned in the adaptations of pattern, in the changes in temper, and in the types of metal tool to be designed for that rock. The entire output of a quarry may be improved by a clever quarry smith.

One smith working in the midst of a wide agricultural district may have to do heavy repairs to cultivating machinery and tractors. In another district, where the formation of the land precludes the use of mechanical cultivators, the smith may have to invent the most weird assortment of smaller agricultural machinery. Elsewhere he may get a fair number of horses for shoeing : most hunting estates have their favourite smith for the hunters.

For example, I give an average day's mixed work in our local smithy. I guarantee that this ' day ' was not specially ' picked ', having lived near this forge for years and gone in and out unexpectedly scores of times at different seasons. The day's collection in this list was an ordinary ' pretty busy day '.

This smithy is an average country workshop, consisting of yard, forge, store-sheds, wheelwright and carpenters' shops, and paint sheds. It is situated off a main high road, among agricultural fields, adjacent to several small quarries, farms, and brickworks. Recently a fair amount of building has been done locally. When I drifted in early in the morning, they were already getting on with some work they had in hand. They were wanting to 'stock up' some more shoes, but 'had not had time to start yet'. Some cart wheels had been completed in the wheelwright's shop, and they were heating the forge oven with logs of wood and rolling the cart rims preparatory to shrinking them on. (Details of the rim-making

BELLOWS. TEMPERING PAN.

THE FORGE IN THE SMITHY

Part of the arch was removed to make it possible to get large objects, such as cart wheels, into the fire. The fire itself must be tended according to the job, thus you have a long fire for lengths of metal, or a deep fire for squarer objects. There is much skill in getting an even heat. Along the edge of the hearth lie the various tools in constant use and the small fire shovels. Below, an old iron cauldron is full of water for tempering. The circular bellows are superseding the longer-shaped bellows, though these long bellows continue in use in many forges completely unchanged in design since the fourteenth century. They usually have a lower handle than the upright makes, though the action is pretty much the same.

The advantage of the circular double bellows is that they blow on both up and down strokes, so you get a continuous blast, not a series of puffs. There is often a small wheel, or portable hand forge also, that can be taken outside for use on some immovable job.

Curiously, in the south of Ireland, quite small cottages and earth-floored cabins have a little circular blow wheel fitted under the hearth to blow up the peat fire. They are sold at local ironmongers and fitted by the people for themselves.

and shrinking are given later, page 155.) As this job was being done, three large cart horses came in to be shod (Illus. 36). While the doorman was preparing these, the smith was called out to consult about a luggage grid to be made and fixed on top of a saloon car. The saloon car framework was strong and the smith, after examining the structure, thought it would take a rack about 4 inches high, but that for the sake of strength and rigidity, it had better be made all in one piece, so as to distribute the weight over the entire top of the car : the measurements for position of the securing struts were then taken with much exactitude. A suitable rod was selected from the shed, and he commenced cutting it till the doorman was ready for him with the prepared horses. After shoeing the horses, the cart tyre was continued, till interrupted by the local transport bus, which came in with a corroded accumulator-carrier. The garage said it required entirely new iron supports. This necessitated the wheeling of the acetylene welding cylinders out into the yard. The smith then lay sweating on his back under the bus with his green glasses on, and the welding blow-pipe in his hand, when four more shoeing horses walked in over his legs (Illus. 35).

Meanwhile, a heavy road lorry rolled up outside, with its safety rails requiring new iron supports. The smith takes a look at the waiting horses, see that one of them must have a specially .made shoe which is going to take some time to fit, and decides to finish the work on the bus first, as that is wanted to fetch the children back from school at 12.

Twelve o'clock is dinner time and everyone needs it.

After dinner the cart tyres are completed, and are put to heat up in the oven. The wheelwright and his department are disputing for yard space with the mended bus, a broken wheelbarrow wanting an iron splint down its leg and a new wheel, a potato plough with a broken tine, and a farm boy with a set of sheep shears for re-grinding ' but they'd do at the end of the week '.

When the tyres are red hot, all hands are called to the yard to help hooping. The photographs (Illus. 37 and 38) are of the tyre shrinking done that day. The wheelwright had been waiting for them to get the job done so that ' his wagon ' could be ' got off '. Meanwhile he had been putting a new tool-box under the car, because ' the last tool-box was about worn through ', and he ' might as well do it now as have it come back to be done in three months' time '. The carpenter's department had also made a coffin, and booked up to re-turn a mangle roller. And as it was June, the smithy bees swarmed in the middle of the afternoon. I *think* that was all that happened that day.

Some seasons they would be busier, some days, if they were slack, they would get on with their own work, stocking up

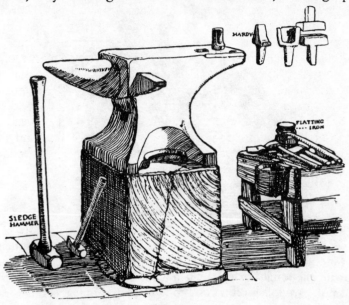

THE ANVIL

The anvil rests on a solid block of wood, roughly shaped from the tree. Quite two-thirds of the block are plain unshapen tree, buried below the ground. In many cases the old smiths probably set up their anvil on the stump of a tree still growing, clearing the roots to get a level approach ; the main thing is to ensure absolute firmness. The three holes on the anvil socket various cutters. On the right the small low table carries the tools, and a pannikin of water, useful for a quick immersion of tool heads. On the left the length of the sledge-hammer, and the average tool, are compared with the average height of the anvil.

repairing tools and so on. But this is a fair example of the variety of jobs a country smith is called upon to undertake.

§ 3. *The Shoeing Smith and his Doorman*

It is almost impossible to give an average time for shoeing a horse ; it can be either very quick or very slow. It is quickest if the horse comes regularly to the smithy, has a perfectly normal hoof, on which the old shoe has not been worn unduly long, and if the smith has, in stock, a shoe suitable

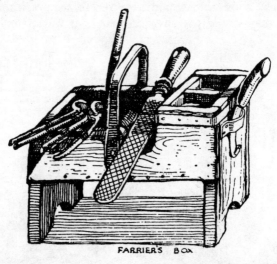

FARRIER'S BOX

for the horse, that is, with the minimum of adjustment (another point in favour of the same farrier continuing with the same horses). Then a capable doorman and skilful smith can polish off a good job 'almost as quick as the horse can change feet', as the saying goes. But supposing a horse has a damaged foot, or some illness or idiosyncrasy to be cured, then the smith may have to make a completely new shoe, working from the bar, and its planning and fitting may take an hour or more. For a good smith

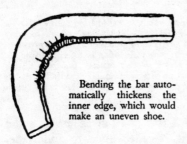

Bending the bar automatically thickens the inner edge, which would make an uneven shoe.

won't 'call it a job' until he's 'got a shoe on' that will be comfortable for the horse, and do the fault some good; nor till he has arranged with the owner when to bring the shoe back to be changed or renewed. It may even be necessary to treat the hoof, and as it is not always possible either to do

148

this at the farm or to return the horse without a shoe along the high road, most country farriers have a paddock or some place where a horse can stay overnight.

Horseshoes are bought ready made in many sizes and shapes, wholesale, but most smiths 'stock up' their own 'specials' in any spare time they have. The 'lengths' are cut off from the bar, till a good pile are lying on the floor.

--- LEVELLING A SHOE -----

These are then gathered into a bundle, up-ended, and picked off in pairs, large or small sizes. They are then forged. Of old, this forging necessitated considerable hammering. Now, the metal can be purchased in different varieties, so the forging is more for shaping than for temper. For example, the bending of the bar of metal naturally tends to thicken the inner curve, and this thickening has to be corrected: also the shoe must be 'true'd up' to get a perfectly level foot and ground surface. The ends of the shoe are rounded in the heel crop-per (see illustration), one of the cutters which fit into the square hole in the anvil. All shoes vary much in make, according to the variety of horse and variety of work and also have to be adapted to the actual idiosyncrasies of each animal.

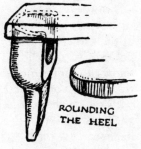

ROUNDING
THE HEEL

The shaping of the shoe is completed by piercing the nail-holes. The nail-holes are not simple holes right through the shoe: if they were, the shoe would fall off at once. A horseshoe nail is very subtly shaped to a wedge formation, so that under continuous wear, it still continues to hold in its socket (see Horseshoe Nails, p. 154). The nail and its socket are a small and most important detail.

149

The calkins and other shapings of the various trade shoes are very technical, but the drawings give a few of the many different types of shoe and show the exactitude required in this quite ordinary country job.

THE SMITH'S DOORMAN receives the horses, tethers them, and has a good look at them to see how they stand ; quietly enough if they are experienced horses, and friendly. As a rule the smith has a good look at the old shoe before he removes it.

The smith taps lightly to tell the horse which foot he wants, and then, sliding his hand down under, the horse's foot obediently lifts to his knee. The smith examines the old shoe in position, notices the wear, and makes a mental note

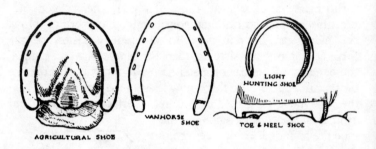

AGRICULTURAL SHOE

VANHORSE SHOE

LIGHT HUNTING SHOE

TOE & HEEL SHOE

of the growth of horn and how long it has stood, also of the position of the clip. If the horse is a 'regular', the smith knows the idiosyncrasies of that particular horse as well as the man that brought it. If there is anything amiss, he consults with the owner and decides on a remedy. Then the doorman, with his buff (hand anvil or buffer to strike against) and hammer cuts off the clinched ends of the nails so that they can be drawn out. He then takes his pliers and pulls lightly, lifting at the heel, first outside, then inside, till the shoe is drawn down from the hoof (see drawings opposite) ; a rap, and the shoe slips close again, leaving the nail heads exposed. The pliers pull out the nails, and then with an upwards and side-ways pull the old shoe lifts off. The foot is then examined carefully, mud and dirt removed, and the sole cleaned over with the back of a knife, so that it can be properly investigated before cutting. The various defects to be looked for are highly technical.

The smith's object is to keep the foot normal in shape,

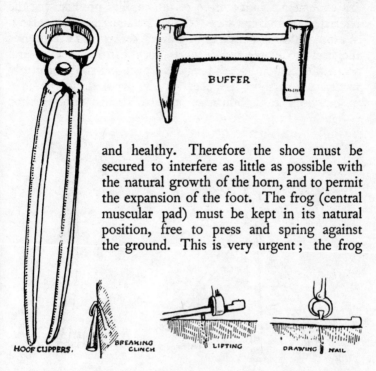

BUFFER

and healthy. Therefore the shoe must be secured to interfere as little as possible with the natural growth of the horn, and to permit the expansion of the foot. The frog (central muscular pad) must be kept in its natural position, free to press and spring against the ground. This is very urgent; the frog

HOOF CLIPPERS.   BREAKING CLINCH   LIFTING   DRAWING NAIL

supports deep-seated muscular action and induces a valvular movement of the blood circulating in the tissues. Thus a clever farrier will spare no trouble to shape and induce

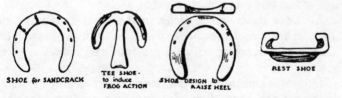

SHOE for SANDCRACK   TEE SHOE. to induce FROG ACTION   SHOE DESIGN to RAISE HEEL   REST SHOE

a growth of horn that, despite the shoe, brings this frog into its full natural action. This is one of the disputed high points of the calkin shoes, fitted to the heavy Clydesdales to give them a grip on the cobbles of their shed and quay-yards. All high shoes lift the sole and frog off the ground, and only the extremely rough surface of the cobbles provides for the natural use and massage of the unnaturally raised frog. Wasting (through disuse) of this muscular pad also causes the heels to close, and in every way damages the natural shape of the foot. The sketches on page 151 show a cleverly made surgical shoe designed to press on the frog, cut short at the heels to let them expand; and also another shoe

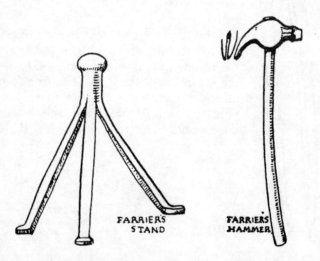

FARRIERS
STAND

FARRIERS
HAMMER

to raise the heel, another to take the weight off a crack; others for the horse that clicks, retarding the over-reaching hind foot and designing the fore-foot shoe to get it clear of the ground as quickly as possible.

All these shoes are examples of the smith's art in making a horseshoe; and the shoes, when made, must be watched, altered, and renewed to help the foot to regain its normal shape.

Doorman and smith need to work together very accurately. It is necessary, in a forge, to have a competent doorman; a good smith means a good doorman, and the doorman can do much to the credit of the smith. It is quite wrong to think of the doorman as merely a halter, bellows, and labourer; he is a very skilled member of the forge, and as between doctor and trained nurse, the longer they have worked together, the easier for both, and the better for the patient.

The doorman having prepared the horse, the smith fits the shoe. The burning is not excessive, and the shoe is not secured till the smith is satisfied it is a good fit and well shaped (Illus. 31, 32, 33, and 34).

§ 4. *Horseshoe Nails*

Horseshoe nails are an excellent example of detailed country work, and were the last nails to be made by machinery. The metal must be extremely hard and yet not brittle. Till recent years they were all forged by a comparatively small section of skilled specialists, so specialized that it is recorded that at the general Nail-makers' Dinners, the makers of horseshoe nails always dined at a table set apart. Nowadays horseshoe nails are chiefly machine-made, and bought by the country smith direct from the wholesaler.

The ordinary nail with a separate head would not do for a horseshoe, because the head is continually wearing off. Sunk sockets in the shoe are still sometimes used (especially in Scotland) to protect the nail head, but they only postpone the difficulty. The nail requires very special shaping to enable it to hold for an appreciable depth.

The metal must be of a consistency to wear equally with the shoe, otherwise the horse would soon stand on six exposed nail points, or the nails would be driven deeper through the hoof at each step. The metal must also be tractable to bend and clinch.

Research was undertaken in the various metal ores by the nail makers; for a time Sweden produced the best quality metal, but now we believe that most good nails come from a North of England firm.

The slant of the wedge-shaped head must be exact enough to permit of expansion under driving to make it not only fit, but perfectly adapted to the nail hole. The detail of pritcheling out the hole for the nail, despite the heavy material and apparent roughness of tool, must be extremely delicately executed. Supposing the wedged nail head sinks into the hole a fraction too deeply, it cannot be driven in accurately. If it stands up a fraction too much, it will be raised above the level of the shoe. And in correcting a nail hole, the hole must be *re-pritcheled from the top side only*. The more obvious action of clearing the hole from the other side would leave a space and clip the nail at a breaking point within the thickness of the shoe (see diagram), so that when the shoe had worn to this point the nail would no longer hold the shoe. The good smith makes so slight a movement in sliding the nails into their sockets and glancing at them before approving the shoe, that it would pass unnoticed in the routine of his action. Yet to omit that one small point in his work would be considered a serious error by any capable smith. People seldom realize that under a mass of heavy work there is such delicate detail of a countryman's skill.

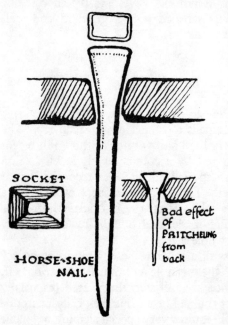

SOCKET

HORSE-SHOE NAIL.

Bad effect of PRITCHELING from back

§ 5. *Shrinking on a Cart-wheel Tyre*

This is a small, representative smith job. All iron cart tyres are shrunk on to their wooden wheels. A round iron wheel table is to be seen outside most country smithies, and

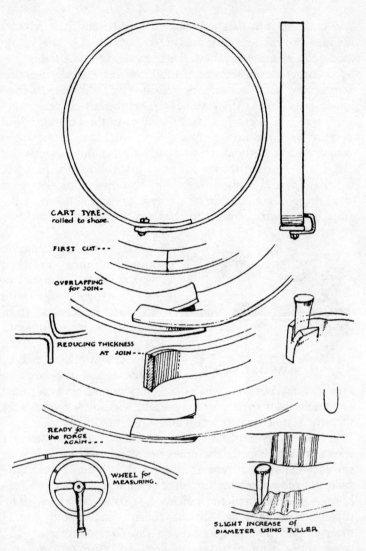

CART TYRE-
rolled to shape.

FIRST CUT---

OVERLAPPING
for JOIN-

REDUCING THICKNESS
AT JOIN---

READY for
the FORGE
AGAIN---

WHEEL for
MEASURING.

SLIGHT INCREASE of
DIAMETER USING FULLER

often you will find a circular grid, somewhat like a skeleton
cart wheel in iron. This holds the tyres raised over the fire,
if the smith heats them horizontally. But, as it is very necessary
for the tyre to be heated evenly, and as an even circular fire is
rather difficult to manage, many tyres are heated in an upright

oven, where iron bars are shoved through and it is possible to turn the tyres against the red hot walls. On the whole, you get a more even heat in the upright oven. As the smithy is adjacent to the wheelwright's shop, wood is usually plentiful for firing. In the oven described here, old sleepers from the railway are burnt, as well as the carpenter's chips.

A cart tyre is cut to shape as shown in the diagram. The bending is done in a simple rolling machine. The size is measured as exactly as possible, and the welding together of the join is done in the forge. The diagram shows the process : a half-thickness of metal is cut off to prevent thickening at the join, and that section of the tyre goes in and out of the fire three or four times in between the different welding processes.

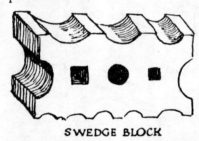

SWEDGE BLOCK

When welded, the inside measure of the tyre is taken very carefully. The amount of shrinkage varies with the size of the wheel from as much as three inches to as little as half an inch. Slight adjustments can be made to get the wheel a little larger, but on the whole, this simple job is most accurately carried out, with no more complicated apparatus than the simple tracing wheel and a lump of chalk.

When the tyre is heated to its full expansion, the cart wheel is laid perfectly flat upon the stand and three or four spikes are lightly driven in around the circumference (the spikes of old rasps from the forge serve very well). The red-hot rim is then lifted and quickly laid over the wheel. The spikes are instantly and simultaneously knocked out, and as the rim falls into position round the circumference of the wheel water is poured over, cooling and shrinking it to grip the wood. Though extremely simple, it is a ticklish job and calls for good team work in the yard, for the smith's red-hot iron must not rest an instant on the wheelwright's wooden wheel, or it would burn the wood. Also the rim must lie perfectly true and level with the wheel or it would never turn truly. The shrinkage must be uniform and the water lavishly poured on till the wheel is tightly gripped by the fully shrunk tyre (Illus. 37 and 38).

The whole process barely takes three minutes, and permits of no corrections if there is a slip.

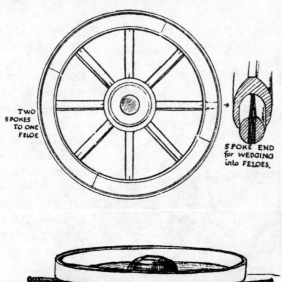

TWO SPOKES TO ONE FELOE

SPOKE END for WEDGING into FELOES.

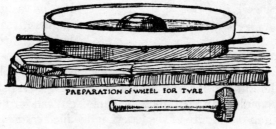

PREPARATION of WHEEL FOR TYRE

§ 6. Buisting Irons

> There must be marking irons for our beasts
> And tools for to gelde and clip and shear.
>
> *Palladius*

'Buisting', or marking irons, are one of the small jobs of the local smith. They are usually simple, of two or three initials, and they last a lifetime (or several lifetimes if the same family, with the same initials, continue the same sheep walk).

Sheep are ear-marked according to their pedigree, and this mark, of course, lasts for life, whereas buisting is a seasonal mark that only lasts during the growth of one fleece. Thus it is an ownership mark for grazing or sorting. As a rule, the ear-marks will tell you of the sheep's family history—as to whether she is a ewe of twins and personal matters of that sort. The buisting tells of her owner. The sheep are marked after clipping and before turning out for grazing. A small iron pot full of pitch and tar melting on a rock-built fireplace, and the buisting irons are part of the equipment of every sheep shearing.

TAR SKILLET.

BUISTING IRON

If you see a sheep farm to let, or taken, you will usually see the initial irons in the local smith's shop. As shown by the example illustrated, these buisting irons call for both good design in the pattern of the letters and skilled craftsmanship in their construction from iron plate and bar.

§ 7. *The Tinker*

Tan ran tan ! tan ran tan lan !
Have ye tin pots ? kettles or cans ?
Coppers to solde ? Brass-s-s-s Pan-n-n-s !
Tinker's Cry

Despite the mass production of tin-ware, a great many tinkers still ply their trade in country places. In the north especially, you get numbers of wandering tinkers who work their way from one district to another. Of necessity they undertake repairs quite as often as they make new goods, and the competent travelling tinker knows his district so well that he can tell you to an inch the amount of wire required for the edging of a milk pan fifty miles away. These folk depend on their wits as much as on their skill, and are clever to notice new makes of stoves or milk separating machines; they will, on one journey, perhaps make a Dutch oven to fit in front of a fireplace, or a splasher, or some gadget, ' as a free gift,' for some housewife, knowing that if she finds it useful, there will be half a dozen other jobs next time.

In certain districts special jobs keep regular tinkers occupied in making some particular form of tin-ware. For example, in a small mining village near us, a workshop employing three or four men (and the usual boy) is kept busy turning out the three kinds of colliers' pannikins. Occasionally they have an order for cans of another sort, but for the most part they go on pretty regularly turning out the perfectly plain round tin pans with a fitted lid for carrying the miners' dinner, the oval bottle-shaped tin for the cold tea, and the queer expanding horseshoe-shaped sandwich case, which keeps the dust out and the gravy in, and is the special case liked by all miners.

These articles are very strong and made for hard wear and are therefore comparatively expensive. The tinker workshops buy up old metal teapots and such things to make their own solder, and in winter when the iron stove is going full blast, and the winter sunlight flickers over the shining tin, it is a very pleasant workshop, though it gets pretty hot in summer because of its iron roof.

In real country places, where open fires and wood and peat are burnt, the heavy iron pots and pans last for years and the tin-smith's job most commonly consists in replacing new tin lids to fit old pans, and dripping-tins and bake-pans to fit old-fashioned ovens, and there is not the same demand for light tin and aluminium ware as in districts where gas and electricity are installed.

But for examples of general tin-smiths' work, I will describe some objects made by an ordinary travelling tinker. This man was working in the North; he travelled a very wide agricultural district where they burnt peat, getting his supplies

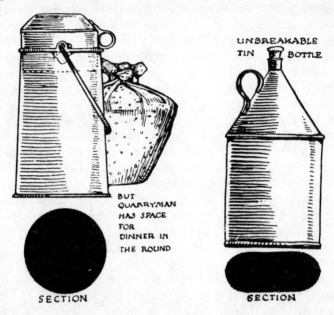

BUT
QUARRYMAN
HAS SPACE
FOR
DINNER IN
THE ROUND

SECTION

UNBREAKABLE
TIN BOTTLE

SECTION

wholesale on his occasional visits to Glasgow and sending them by rail to depots up country, to be conveniently collected by him later. He travelled by cart—that is, he had no van, but carried a tent which he pitched at will in some district that he could work on circuit with the light cart for a week or a month and then move on to the next place. He worked to order, relying on making and selling things in the district, unlike some van-dwelling tinkers who make a quantity of stock and then rely on selling it off as they clatter along (Illus. 39).

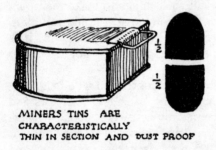

MINERS TINS ARE
CHARACTERISTICALLY
THIN IN SECTION AND DUST PROOF

$\frac{1}{2}$

$\frac{1}{2}$

I located this particular tinker while lying half asleep at noon-day on the sandy edge of a loch in Skye. I had been listening to a faint tapping that I took to be birds shell fishing or seaweed drying in the sun. When I stood up, I could hear nothing, only the wind stirring the loch and the cries of the gulls following the rising tide. But a foot from the hot earth I could still hear it, tack-tack, tack-tack, and all of a solid 2 miles across the foot-hills slowly I tracked the sound till I found the tinker's tent about half a mile from the village.

TRAVELLING
TINKER'S
ANVIL

His bench was the turf, his anvil an ordinary tin-smith's light anvil, but adapted for him by some smith so that the long pointed leg would drive some depth into the earth. He had a couple of wooden boxes that fitted a-longside the cart, and these when lowered fitted either side of the tent (Illus. 40), holding down the canvas. The iron half-hoop that held the cover over the cart was now driven into the turf for the tent. Beyond the fact that he kept his tin in small sheets packed flat along the bed of the cart, and that his coils of wire were of smaller diameter, there was no difference between his small tent workshop and that of any other stay-at-home smith.

He made me the rather deep-lidded cooking pot that I asked for. He used, by request, a fairly heavy sheet tin, a small rimming tool, and the ordinary mallet and shears of tin-smith's use. A small hole in the anvil accommodated the handle of the shears. He used a Primus for soldering : these stoves are in common use now among travelling tinkers, replacing bellows and brazier. He marked the sheet, bent up the shape, secured the seam, bent over the wire of the rim, bottomed and handled and completed the job with the

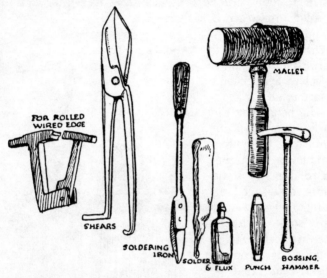

simplest apparatus and the greatest skill. It took him about half an hour and it still wears very well.

Other things made by the travelling tinker include baths for babies, wash-basins and wash-up bowls, cake tins, flat tins for ovens, toasting tins, toasting forks (of twisted wire, sometimes with wooden handles), milk pans for setting the cream to rise, milking pails with specially narrow tops for milking in the meadows and carrying, baking tins, dripping tins, small Dutch ovens to stand before the fire, cowls to fit the chimney and pull down over the wood or peats to make them burn up, and similar cowls for coal fireplaces. He does odd jobs on the mudguards of bicycles and perambulators, tun-dishes, and funnels ; sometimes he makes the special oil cans for cars or motor tractors in the fields. He usually makes, too, a few ' regular line ' articles likely to sell locally, and hawks them about the village, booking orders for special sizes, and finding out whether it is likely to be a good district or not. They usually work in tin but occasionally use copper.

One tinker, whom I rebuked for his tardiness in returning my tea kettles, said, reproachfully, ' I've been awaiting copper-bottoming them, mum.' (It wasn't until I tried to repeat the remark that I realized he was a master of elocution.)

To-day there are fewer travelling tinkers than there used to be. More's the pity : they did good work and lived a healthy, happy life till the Vagrant Laws herded them into crowded town tenements—where they usually died. The towns are too full already. Why drive more country people into them ?

CHAPTER FIVE : BRICKS AND POTTERY

§ 1. *Bricks*

PEOPLE TALK of ' old brickwork ', but, except for Roman brick, no brick found in England is of very ancient manufacture. Early English building material (other than stone and timber used for fortifications) was wattle and daub, or varieties of the ' Cob ' building of Devon. Wattle and daub, described on p. 100, was by no means the rough structure it is popularly supposed to be. Provided the stakes were well driven, the cross bars of the wattling firm, and the clay and binding admixture well worked in, a wall so built was an extremely solid structure.

After the Romans left, the brick-making they had originated fell into disuse, and their brick buildings were quarried for material, so many old English walls are built of stone mixed with Roman brick. It is believed that good bricks were made in Suffolk as early as the thirteenth century for we find mention in the country of brick buildings of that date. Hampton Court shows to what perfection brick-making had been brought in Henry VIII's time, and very many old Elizabethan houses and garden walls of brick date from that period. Many houses, now apparently black and white, are of brick and timber, but the brickwork has been covered with plaster and whiting. Originally this can have been done to keep out the weather, but in some cases it is as great a vandalism as the thick incrustations of tar or glazed black paint upon the mellow timber. The date of bricks can be judged approximately by their size. The earlier ones were irregular in shape, and often very thin: after 1625, bricks were standardized, 9 by $4\frac{1}{2}$ by 3 in.

The building boom has stimulated activity in the brick-fields all over the country. Some few small ones are working again (that is if they had merely been ' closed down ' and not ' worked out '); but with the improved facilities for transport the tendency seems to be to enlarge the working brickfields rather than to develop new ones, or reopen the old.

We find many variations in bricks. In some places there is a good yellow clay which makes fine firebricks; the quality of this depends on more than the colouring. Oxide of iron is an important ingredient in brick and the method of making and firing also account for its colouring. Other bricks (for example, those made in Kent) are a pale primrose yellow, but are not firebrick. Cambridgeshire made very pleasant yellow bricks and tiles.

Details of the constituents of brick are technically beyond the scope of these notes. Briefly, brick is a mixture of clay, sandy earths, and firing substances. The ingredients brought to the brickyard to improve the existing clay vary with each locality, some 'fat' clays (the word is technical and well describes the rich pudding-like clay) will bear a large admixture of sand, often carted from a distance. Sometimes chalk is used in a mixture of clay and chalk and sand. In these districts a sort of brick soup is made of rubbed-down chalk and rubbed-up clay, both sufficiently liquid to be puddled together and thickened off with other ingredients. The quality of the sand itself must be studied: small rounded pebbles (if sieved sufficiently finely) may be good, but any limestone gravel is fatal, for it burns to quicklime. On the other hand, river-sand containing organic matter, or sea sands containing salt, must be investigated carefully, as some sands 'rot' the earth. Since the local ingredients are so varied we find highly localized varieties of brick in the smaller English brickworks.

The brick substance is puddled, and when of the right consistency, is moulded into bricks. These are quite often made by hand, as the process is simple and quick. Bricks are dried in sheds which stand in long lines, no more than low shelters under which rest the long lines of brick-laden planks, sheltered from the rain, but open to the drying wind. The large brickworks sometimes construct the flues in their kilns to run under the brick-drying sheds, and hasten the process, but in the real country brickfields, brick-making at this stage depends almost more than any other country job on the weather.

After ' drying out ', the bricks are burnt either in clamps or kilns. Kilns vary, the most usual type being round in shape.

The chief difference between a pottery kiln and a brick kiln is that the pottery kiln is more like an oven, in which the pots are piled for cooking, while in the brick kiln, the bricks themselves form part of its inner construction. The simplest kiln still in general use is fired from below and the heat burns up and out at the top, but some kilns are made to burn downwards; there is wide variation, both in building and in fuelling.

Plain ' standard brick' forms the largest output, but every brickyard naturally produces many special ' pattern bricks' to match its own stock. The number of differently shaped bricks required for one simple building would surprise a layman : and as it is a characteristic of brick building to use small and rather interesting ' finishes', you will get perhaps twenty or thirty different types of bricks, small and large, all employed in connection with one standard building brick. I will not involve the reader in technicalities, but it is obvious there must be special flattened-out bricks for window ledges, that where the window ledge turns up at the side of the window a shaped corner-brick is needed and that these must be made for right- and left-hand corners. For the string courses there must be the smaller sectional brick and a variation of this type in the two sorts of corner-bricks—for turning in and for turning out. Around cornices, or where the bricks are set inwards, there must be the half-round bricks, to slope the rain downwards, and this half-round brick must turn the segment of a circle of 45 degrees at the corner of a building ; above doors, along the ledges of roofs and gables, above and below ground different types of bricks are used, and for each there must be a right hand and a left hand, so that you will find twenty or forty moulds of brick all employed in one simple home.

These variously shaped bricks have definite names, according to their position in the building scheme (and right or left hand, upper or under and so on). Besides these technical *shape* names, there are the names given to the bricks themselves according to their variety and texture, faults of firing, or moulding.

Practically every type of brick has its own name and that

name is different in every brickyard. One interesting name was ' Slippers '. I naturally took it to mean they had ' slipped ' in making, but no—they were ' shippers ', so called because they used to be sold off for ballast for the ships that brought the sand or the chalk to the pit. ' Burrs ' was an obvious name for bricks that had roughed and stuck together, and ' chuffs ' or ' chucks ' were sound but unsightly throw-outs that could be used underground.

§ 2. *Making Kentish Brick*

As always, it is difficult to decide, on the most representative example of a thing made in many parts of the country I choose Kentish brick, not only for its historical interest and its excellent quality, but because this type of brick is made, both in large quantities by advanced methods with machinery and also in the smaller Kentish brickfields where some of the simplest traditional workings may still be found.

It is believed there were brickworks in the Kent district in the thirteenth century, or even earlier, and tiles were produced at the same date. In 1374 local kilns turned out 190,000 tiles, and down the centuries, accounts from monastic institutions show a steady growth in the size and number of tiles.

In Kent the brick clay or earth, generally called by the workers ' the dutt ' (the dirt), is found with clay, and there is plenty of sand near by. Every brickyard varies in methods, but on the whole the procedure is to beat up the clay, and beat up the dirt, till both are sufficiently liquid to be pumped into the ' wash-backs ' or great pans in which, after stirring, they are left to settle.

The firing principle is worthy of note ; here it is ash and coal dust transported by rail from London's backyards (one worker described this to me as being ' full of the most suitable impurities '). This combustible element is incorporated in the texture of the brick, so that, as the clamp hands say, Kentish brick is good because ' it burns itself right through '.

The finished brick is a pleasant cream colour, in appearance much like the whitish, and cream and brown shaded old tiles of the Cambridgeshire district (some of the good tile works

167

in East Anglia have unfortunately closed down). The fine cream-coloured buildings of Kentish brick that have stood for centuries around the south-east of London testify to the bricks' enduring qualities.

There is an excellent brickfield situated on the mainland near the island of Sheppey. It is set fairly close to a small river estuary, from which sand can be obtained, and down which the bricks can be exported. The puddling is done in square open pits; there are kilns further inland for some of the work, but nearer the road great quantities of brick are built and burnt in primitive clamps. I will try to describe the work in detail.

The long wooden drying sheds are near by, and the dark weathered wood looks very pleasant between the white brick clamps, the green fields, and the red earth. There are rail lines with iron trucks, and a little machinery. The work is carried out very simply and the bricks are of good quality.

The seasonal nature of brick-making requires frequent visits to the same yard throughout a year if the work is to be summarized. In October, the mixed clay and earth are being pumped along heavy wooden culverts to fill the wash-backs, which are large square shallow ponds in which the mixture must stand and set for three to six months. Perhaps the damp grey drizzling day of my visit, the mist hanging over the river, and the drip of autumn somehow influenced the slow tempo of this part of the process. A silence and haze hung over the landscape; a steamer in the estuary hooted mournfully and the thick semi-liquid brick flapped and plopped accompaniment. The sides of the square pans are from 2 to 6 feet high, and around the four sides is a small muddy path. Along, overhead, crossing the pool in all directions, are rickety wooden aqueducts leading from the pumping-house. Standing on the edge of the pond, wrapped in silence, I watched mud. Overhead a man came, walking slowly along the planks; slowly he walked along, stooping at intervals to lift the box-like ducts so that the heavy mud should fall evenly into the pond. There were no elaborate valves or sluices. By the simplest arrangement, the lifting down of one box end from above or below the next, left the brick soup free to flow on or free to fall heavily down into the pond below (Illus. 41).

There would be a dull wooden bump and a slow slobbering sound, and the mud flood would begin to flow down; then very slowly the man above straightened his back, carefully placed his two feet on the narrow plank, and arranged his arms akimbo on the wooden hand rail. Slowly he brought a dreamy gaze down from the boundless horizon to the shapeless mass below. He spat judiciously. He watched . . . he spat.

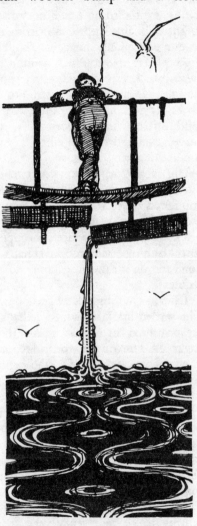

Grotesquely distorted in reflection, the wavering upside-down-himself that wobbled outwards to the brim of the pond but emphasized his immobile stance above. Tiny dusty bubbles broke muddily as the dull spreading circles wavered outwards. A grunt overhead, of wood and of human simultaneously, the duct was replaced, the flow stopped, and the brooding workman moved off heavily down another plank way, the supporting poles curving sympathetically below his pondering tread. Doubtless he went to brood upon other ponds.

The heavy mixture in the pool subsided slowly. A thin watery gloom filmed over the top. A brick pond on a wet day is a depressing spot. . . . There was a faint sigh behind me. It expressed no haste, no sorrow : only it reached me as might a lullaby lapped in porridge and the reflections began to move again. A quiet man, with mud-coloured clothes and a peaceful expression, was waiting to pass. A long pole went from him horizontally into the pond. Slowly the pole slid through his large hands down into the pool and slowly passed back. It was difficult to know whether he stirred mud or mud stirred him. Deep-sunk, in slow detachment and peace, they stirred together. I watched him for a long time. . . .

His is not an arduous occupation. He follows as the moving mass is poured out. He stirs his mud until the appointed time, when the liquid may cease to move and lie still to solidify. As I turned and slowly left that October brick pond his voice dropped down through the soft lapping silence : ' That 'uns done . . . that 'un don't need stir'un no more now. . . . ' It sounded like a requiem.

Later, in spring, there is a new spirit in the brickfields. The settled brick is ' soiled ', that is to say the fire and ashes are scattered broadcast upon the soft surface much as sugar is strewn on porridge, and a consistency much like porridge results, for the whole is stirred together by spade and boot and pudmill (mixing the PUD, as its name suggests) (Illus. 42).

From April to September the lumps of brick clay from the pudmill are cut, and set to dry. They are laid out in the long drying sheds and the drying process takes from fourteen days to a month according to the weather, but it cannot be done at all in frosty weather, as frost would expand the moisture in the brick and spoil it.

The method varies so much in each brickyard that it is difficult to be representative. In one shed there seemed to be about six workers, and their names indicated their share of the job : ' moulder,' ' temperer,' ' off-bearer,' ' Flattey ' (or ' Walker Flattey ', the man who walks about and sands the work), 'pushey' and 'barrow loader'. Between these workers the lump of clay is moulded, and the bricks are piled up, and then taken off and placed more widely apart for drying. The fact that this monotonous job is shared by so many workers is a good example of the instinctive way in which sensible country people can, if desirable, specialize and divide routine jobs to everyone's ultimate gain and the more satisfactory completion of the work.

The dried bricks are now ready for 'crowding', the word used for stacking them up and wheeling them to the kiln or clamp. Finally, towards the end of the summer, and overlapping with the beginning of the winter work, comes the firing of the bricks.

The day I went round to see the firing was a blue gold autumn day. The brickfield looked like a film-set showing Assyrian antiquities. Long serried walls, mountains of cream-white brick, narrow oblong avenues, squared the shadows of sunlight. The immense size of the clamps dwarfed the workers clambering over them. A clamp is really a small hill of solid brick built on the level ground : foundations of burnt bricks are laid in open work and the 'breeze' or firing material is laid skilfully in flues over which are piled the bricks, sometimes to a height of fourteen feet. When they begin to burn, great screens are used much as the charcoal burner uses his screens to fan his hearth (see p. 48). These enormous screens of wattle or sailcloth are shifted round like great black banners to catch the air, and the blue smoke and the gold flame pour down the wind.

When the clamp has burnt out it is left to cool, and then taken down. Skilled workers climb up along the clamp, sorting the bricks as they pass them down. The old covering bricks that were used outside the clamp are thrown into one pile, good bricks to another, discoloured bricks that can be used underground or for inside work to yet another; 'firsters,' bricks of the best quality burnt to a fine even texture, are given a special pile.

And so the brick mountains are sorted out into small hills and in due course the small hills sent away to their appointed places.

§ 3. *Pottery*

If thou hast a piece of earthenware, consider that it be a piece of earthenware and by consequence very easy and obnoxious to be broken. Be not therefore so void of reason as to be angry or grieved when this comes to pass.

Epictetus, ex-slave.

Pottery is one of the most interesting examples of localized design. Its weight and brittleness mean that one design will remain long unaltered in one locality. It is the basic craft of an agricultural people. Nomads must travel light, and so bags and skins and light metal ware take the place of pottery among hunting tribes and wanderers, but when a people settle and take root, we find the primitive funeral urn and the great corn jars. In England very fine china and porcelain have been manufactured for a long time, but only in certain localities. China was sufficiently valuable to be worth its early importation from overseas, so fine porcelain and china were brought here from the ends of the earth; but the plain, thick, common pottery of everyday use was used locally, and bears unmistakably the stamp of its locality. Its character is based on the local clay, and it is designed according to local ideals and to meet local needs.

Originally there were many potteries scattered about England; wherever there was a good clay bed there would usually be a local potter. Nowadays, despite the fact that the industry is centralized in 'The Potteries' and despite increased facilities for transport, there are a surprising number of small country potteries still at work. I am not counting the 'arts and crafts' and 'studio' potteries, or those turning out decorated 'Gift Ware' and helped in the holiday season by transported mass-produced goods: I mean those potteries occupied in turning out practical everyday goods for their local market. They are little known, because, as a rule, the goods are made for small local dealers and local markets.

Up to about the eighteenth century these potteries distributed their wares within an approximate 50 mile radius. The distance of course varied, but by pack pony not more than 50 miles could be done conveniently in one expedition—say 25 miles delivering the goods, 25 back collecting payment or the exchange goods on the way. Thus 50 miles, in each direction, would give you the same common pottery pieces found in use 100 miles apart. Nowadays these potteries still deliver goods within a 50 mile radius, because that is as far as a lorry can conveniently travel, if it is to deliver its load and return in the day. (Incidentally, the increase in the growth of bracken—'the bracken menace' as they call it—has some connection with the transport of pottery, for these small potteries out in the remoter country places used enormous quantities of bracken for packing. Now that it is no longer cut for that purpose or for bedding for animals, footing stacks, firing ovens, or clamping roots, it is spreading beyond control.)

Some of the fine china that was made for the rich tended to become ornamental rather than useful, but the thick local pottery that was made for everyday use needed no decoration and so the potter could concentrate on making his simple and useful goods of perfect shape. Any ornament he added was for the sheer love of his craftsmanship.

The quality of the pottery depended on the local clay bed ; the type of ware depended on the local demand, and the dishes were designed for the prevailing style of cookery in the district. A wood-burning district needed a different type from a peat- or coal-burning district ; different districts have their own specialities in food and therefore the cooking pots for that speciality will be found in its own particular neighbourhood. Since there was little reason to carry cheap heavy kitchen ware from one district to another, the pottery designs of one district were kept distinct from those of another district. Only when a certain district specialized in something so good that it became really famous did its particular pottery travel further than usual, and the name of its locality might ultimately be given to pots copied from the original shape, but made with slight variation in clay of another district. For example, the old cookery books often mentioned baking in a 'Nottingham' pot, and the 'Lancashire hot-pot' had its pot of special shape (probably to stand on the rotating shelf of the iron ovens). The 'Buckley milk-pudding dish ', the ' Devon milk panshon '—designed through centuries to throw up the cream—and others, these acquired a reputation that sent them far afield, but more often each district localized its pottery, and this localization forms one of the most fascinating branches of English country work to study. The wine jars from the district near Notting-ham are quite different in shape and colour from those in use in the Marches, which are different again from the wine jars of South Wales.

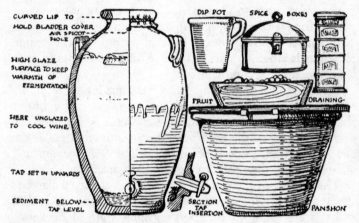

There is much subtilty in making a good wine jar, as shown here; with some other necessities for home-made wine.

Local pottery must not be confused with the pots the gipsies carry round in their carts. From time immemorial these nomads have traded in pottery along the country routes and sometimes no doubt they carry ware from a local pottery to outlying country districts. But as a rule, nowadays they carry the misfirings and throw-outs from the big central potteries: misfirings are dishes and pots that have some defect, and are therefore unfit for regular stock.

§ 4. *Buckley Ware from a Local Pottery*

I choose to describe the Buckley ware as a detailed example of English country pottery, because it happens to be the ordinary local kitchen pottery that I have used all my life.

Some of the larger Buckley works are very nearly 'town industries', but the smaller kilns maintain their reputation for turning out cheap pottery that is simple and good. When wood and peat were used as regular fuel, and mixed dairy and sheep were the staple industry, the designs were particularly well adapted to the needs of the locality. The ware is changing its character, and unfortunately output, with the innovation of coal and gas cooking and the centralizing of milk products; but there is still a fair demand for Buckley pottery by Buckley people and a fairly promising adaptation to new conditions (Illus. 43).

There is the usual radius of distribution by lorry on the main roads and by carts on the small hill roads. The improvement in methods of transport has reduced the amount of

packing needed, though they still use a little bracken. (Rye straw for this purpose is specially grown near some potteries.)

The ware is a deep reddish colour, fairly thick in substance; when glazed, it takes every shade of wallflower colouring, a fine golden brown, rimmed or lined with a 'slip' of primrose yellow. Sometimes this last colour is used for a very simple decoration of swiftly-drawn curves and lines, sometimes for labelling of the dishes. The dripping dishes, for instance, will have 'Beef' or 'Mutton' or 'Pork' writ plain across them, because in a mountain district it is important to keep the common mutton fat separate from the precious beef dripping. In the same way we commend the sense which puts 'Steak and Onions' around one baking dish lest a milk pudding thoughtlessly baked in it later should be haunted by the ghost of the onion. Or a mug may be labelled 'One Pint', or a milk jug 'One Quart', and a frying dish says 'Eggs and Bacon', but the wording is always strictly useful, never inane.

175

These brown pots wherefrom I have carved many a succulent slice and in childhood sopped the gravy, are thrown by hand, on the most simple wheel, and fired in a perfectly simple kiln, and sold at prices which compete reasonably with the specimens of mass production that are poured into the neighbourhood. We like them partly because they are of really good design and partly because we ourselves are very conservative. In small country households dishes are individually known to us; they get broken as all pots may be, and are replaced as most pots must be, but we who do our own cooking learn to know the capacity and capabilities of our cooking-pots very intimately. We do not use ' *a* dish for an apple charlotte ', we use ' *the* dish for the apple charlotte ', because we have learnt from experience that that particular dish is just the right depth to cook it evenly, and just the right size.

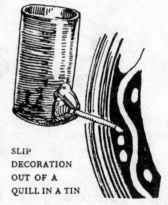

Farmers' wives like the thick deep earthenware dishes for their milk puddings ' because they can stand on the hob all morning ', since the thickness of the pottery distributes the heat very evenly. So that, as far as we are concerned, we always keep up a very definite demand for our local pots.

SLIP DECORATION OUT OF A QUILL IN A TIN

It should be mentioned that at Buckley, which is in the county of Flint, one clay pit is of particular interest as the cupping of the land so tilts that it remains level, no matter how much is drawn out. In the works now they have a motor engine to turn the wheel, which also drags the grinding stone round its trough preparing the slip (a special mixture for lining or decoration). The workers would laugh if you called them artists, yet there is no artist or designer on the premises other than themselves. They make the shapes that they ' think will be most useful ', and they are most useful. If they put on any ' decoration ' they put it where they ' think it would look well ', and it usually does. The concentric circles which make a ' nice finish ' to the thrown jars and bowls are run on the potters wheel. For flat work, the apparatus is no more complicated than a cocoa tin pierced with a goose quill, and through this the ' slip ' is poured out with a judicious hand.

The drawings show the ordinary method of moulding; the clay is spread over the pie-dish mould and trimmed round with a knife exactly as the cook will presently trim the pastry away from the same rim. The output is largely of plain round bowls, wash-basin shape, so much used in the district, deep bread panshons (larger than the usual commercial shape to hold the larger home-made loaves) and the milk pudding dishes (which have a considerable reputation for producing really creamy milk puddings). They make also a fair number of wine jars. But the goods in the yard and store are always pleasantly varied. Sometimes there will be half a dozen large oval panshons, the size

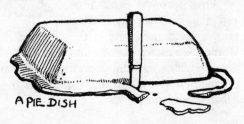

A PIE DISH

of a baby's bath, for curing hams ; sometimes rows of the deep handled pots that we use for bailing jam or fruit messes. Another time there will be a kiln full of hot-pots. (Again notice the local make, for the Lancashire hot-pot must take the long chop bone of a large northern sheep, but those pots that accommodate the tiny Welsh mutton can be a different shape).

Once there was a kiln full of pottery spittoons for public houses : another time they concentrated on jugs, large and small, some with glazing all over, some glazed only on the inside, and the outside left porous with the idea of keeping the contents cooler. Sometimes they would set to work and make a whole lot of heavy bread pans and jars. Once, in an enterprising moment, there was a burst of truly terrible ' ornamental objects ', which petered out in a light-hearted vein of memorial vases.

At the kiln I know best, it is always an interesting speculation as to what particular branch of pottery they will be firing next. As a rule each kiln is packed with more or less the same things.

§ 5. *Some Other Potteries*

Other potteries up and down the country vary considerably from this one I have de-scribed. The Devon pot-teries make the well-known soft porous reddish pottery. Up in the North there are country works where they make the heavy stoneware drinking pots and cattle bins and the stoneware of farm life. Sussex, and some of the Home Counties, have pot-

Solid pot feet to keep wooden furniture off wet stone floors.

teries which were utilitarian in output, but now chiefly make ornamental vases, and mugs and jugs for the new residential urban population.

§ 6. *China*

This should on no account be confused with pottery, earthenware, or stoneware. China and porcelain works are never country occupations now. But the digging of the Cornish china clay, though a huge commercialized industry exporting clay to the East (for weighting silk) and all over the world for many trades, is a very countrified industry, which is, however, carried on, on too large a scale to be described in detail here.

Actually the china clay industry is not very old. William Cookworthy, of Plymouth, seems to have been the earliest worker in 1750, and the first workings were not at St. Austell, but at Tregonning Hill, though Plymouth and Bristol made some of the earliest china. China stone should not be confused with china clay.

The white streams of milk-like water flowing through the green fields of Cornwall are from china clay works. The clay is water washed, and drags and micas and various names are given to the sieving processes that remove the small pebbles, and the clay is subsequently settled in pans as a thick white deposit, to be later packed into barrels for transport.

CHAPTER SIX : LEATHER AND HORN

§ 1. *Country Leather Works*

IN ENGLAND and Wales there are about 36,000 tanners and leather dressers, one-tenth of whom are employers, managers, and foremen. Special tan, lime pit, and yard workers (not simple labourers) number nearly 2,000. The large proportion of managerial posts to skilled workers is accounted for by the nature of the industry and the comparatively numerous small works in the country which have persisted in spite of the growth of large town factories.

The modern tendency for the smaller country works to become localized in one centre and there carried on under factory conditions is nevertheless very noticeable in the leather trade. Improved transport has wiped out many which formerly clustered round the old drover-roads (which are referred to later), but there are still enough to justify this being considered a ' country ' trade. Most modern processes show a series of improvements rather than any direct alteration in method from the days of the independent worker and the country shop.

These country workshops often specialize in one definite leather product, such as roller skins for the colour printer, glove leathers for the home worker, warm slippers and overboots (usually of natural lambswool), bellows leather for the smithy, or oak tan leather for shoes. For all such products there is a small but regular demand.

The large leather factories can experiment and pick and choose among the variety of their ' fancy ' finishes according to changing fashions, but the country workshop must rely on simple processes. The exquisite modern fancy leathers, often made from the skivers (split skins), demand for their success complicated plant, delicate colouring, and ever-new finishes, all possible only to the larger factories. It is the production of chamois, parchment, sheepskins and the like, no less skilful but simpler processes of more unchanging character, which continue as country industries.

I have been intensely interested in contrasting the almost medieval simplicity of some of the earlier leather processes still in use, with the modern complicated methods. There are curious similarities, too. Thus the skinner or flayer at the slaughterhouse, before the skin reaches the leather dresser, if impatient, sometimes uses a knife to sever the skin instead of peeling and stretching by force of elbow and wrist, and this primitive fault is as common a complaint in large city factories as in small country shops. Dye-vats of coopers' work, and mash-sticks and draining-boards unaltered in main design since the Middle Ages are in use both in the city and country works. But the constantly changed electrotype printing plates of photographically reproduced casts of reptile skins used in the large factory would be impossibly expensive for short-time use in a small country shop.

The following notes and photographs are from small works in the country. These can be divided into three main groups : chamois leather, parchment, and the natural wool and sheep-skin products. In the preparation of parchment nothing but quite primitive apparatus is used, with the exception of the large, finely tempered scrapping knife, yet the parchment is of the finest quality and is exported to all parts of the world. Another ' skin ' requiring only skill and simple apparatus and therefore excellently made in the country, is chamois leather. This is dependent to some extent upon geographical conditions, as the quality of the water is important. The following sketches and notes were made chiefly in Cambridgeshire, but the district around Glastonbury has actually given its name to the soft leather products, including natural wool-lined slippers, which are made there.

There are many country shoemakers still working : not only cobblers who mend shoes, but very good craftsmen who build shoes and boots to order. Together with the country saddlers and harness makers, they usually get their leather from a town tannery or import it, though there are several country tan yards still working.

An interesting point is the fact that most leather trades are found along the old drover-routes. In medieval days, the killing of all domestic stock before winter, and the sale of cattle at huge local markets and fairs, made the offal trades concentrate near these definite slaughtering centres. Thus most leather centres can be located at the nearest good water supply at the *end* of the drover routes. An example from the Welsh border country shows the sequence of drover-roads across Wales, the sorting depot at Clun, the selling market on the border; then, below Gloucester, the leather and horn, and below Bristol, the soap works (for the flint, the spongy interior of the horn, was a valuable ingredient in soap making).

I have walked all along these drover roads down to the leather and old horn works in the Severn valley, and the sequence is evident throughout; and it can easily be followed elsewhere.

§ 2. *Chamois Leather*

Leather as maye bee gentle and plyaunte.

(1575)

What is called chamois leather was made in England at a very early date, and the crusaders wore chamois leather suits under their armour. It is, however, questionable whether the genuine chamois ever reached England, sheepskin being used instead of the real chamois skin. In one of the interminable medieval poems, the sheep boasts to the other animals of his skin's value to man.

Sometimes the medieval theories of material were based on very childlike foundations. For example, fat from the swift deer or the limber hare was obviously the fat to use to loosen stiff muscles; those plants that grew in water would probably be good against cold, and so on. Thus as the leather was originally called after the wild and agile chamois, medieval folk thought it a quite reasonable belief to suppose that its suppleness could be transmitted. Similar reasoning is shown in some old English laws for ' poynts ' : the strings, laces, ties, and all the unreliable devices which, for centuries, upheld medieval breeches, were by law made from ' wilde ware ', as being less liable to break under stress. All the same, it was as natural in those days to 'slack off a few poynts' before bending, as it is for a navvy to shift his braces to-day.

The preparation of chamois leather calls for a supply of sheep skins, oil, and special water, obtainable only in certain districts such as in the Eastern counties (Sawston in Cambridgeshire and elsewhere), the Yorkshire dales, and around Glastonbury.

Chamois leather is the inner portion of the skin below the smooth and slightly glazed surface. Originally the surface was sometimes sacrificed for the chamois, or the chamois for the surface; to obtain both, the skins must be split into two layers. In smaller works this used to be done by hand. To-day the same dexterity is used, with swifter and probably better results, in adjusting the long knife blade of a 'splitting machine'.

Once the two layers are obtained by this splitting, the two processes continue separately. The skin may be split so that the surface layer is right for parchment (which is described in the next section) or split in a slightly different manner for skiver, a term used for other varieties of this surface layer. The skiver lends itself to all manner of treatment, dyeing, ornamenting, embossing, and is therefore not often dealt with in the country workshops. But the chamois, or inner skin, which requires a less complicated, though in no way less skilful treatment, is often still prepared in the country.

Theoretically the procedure is to distend the porous character of the skin as much as possible and substitute an insoluble oil filling for the natural soluble grease. To this end the skins are scraped and then soaked in the limewater till clear and white.

They are then repeatedly washed till absolutely neutral and clear of lime. Next, gelatinous dressings are soaked into the lime-distended pores of the skin. Nowadays, even in the smallest workshops, they have heavy automatic presses driven by oil engines or by water power to pound the skins in the thick oil. The medieval recipe asks for 'stock-fish's oil', but to-day Newfoundland cod oil serves. Note that the inspiration to waterproof a material with a fish oil, though medieval in origin and reason, seems a success to-day. The oil, which is driven into the chamois under the beaters, is thin and transparent. The dark orange slime which comes out of

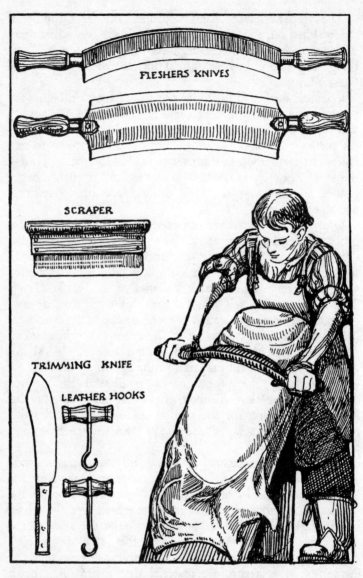

FLESHERS KNIVES

SCRAPER

TRIMMING KNIFE

LEATHER HOOKS

THESE LEATHER HOOKS ARE USED WHEN IT IS IMPOSSIBLE TO LIFT
THE SLIMY WET LIMP SKINS BY HAND.

the skin afterwards, under pressure, is called sod (sodden ?) oil, and is afterwards used for heavy leather tanning.

After the oil penetrates and the soluble grease is removed, the chamois is washed in a hot water vat, subsequently transferred to cold, and the processes of pounding, washing, pressing, and drying are doing now by machinery the work which was originally done by hand, or foot, but not altering the method in any way. Then finally the skins are hung up to dry in an airy shed, when they become quite stiff. They are taken down for finishing to the grinding shop, for grinding wheels now replace the old hand-beating and pummelling. The grinding wheels are made of composition instead of the old natural stone, but this is the only appreciable change in the finishing process. The stretched skins are swept across and across the grindstones till a velvety surface is obtained, probably more quickly than by the older method, but the result in the finished chamois seems identical.

This country trade is one where the modern efficient methods are largely due to the knowledge and skill developed by the old countryman who worked alone, studied his medium, and introduced on his own initiative any process which he thought would be an improvement. Old records show this well. In 1727 we find notes on the method of obtaining the final polish for the finish, which say that it is the pulling across an iron edge ' which opens and softens and makes them (the skins) gentle '. Some workers introduced the oil ' on wool fibres ', making the leather into ' parcels wrapped in wool '. Altogether there were probably many small variations owing to the individual initiative of the country workman.

§ 3. *Parchment*

There is also made of sheepes skin, pylchis wraps and gloves to drive away the cold.

There is also made good parchment to write on. Bookis and quires manifold.

From a thirteenth-century poem referred to on page 182.

Parchment making is one of the oldest crafts we have, and is still carried on practically unchanged. Originally parchment was used for book production, writing, and most purposes for which we should now use paper; and before the introduction of glass both parchment and horn were used in windows. But books were the chief reason for the quantities of parchment made in England. It is curious to think that early book formats (or page sizes) depended on the measurements of the sheep (even more than our writing or drawing paper conforms to those of the ' double elephant '). In Elizabethan days, after the introduction of paper, all sailors and merchants going abroad were exhorted, again and again, to take samples of our English parchment overseas, as ' our workmen make it very well, we have quantities standing, and we need a vent for it '.

Nevertheless, the advent of paper was bound to reduce the output of parchment, though right up to the present day parchment has been the medium of legal usage, all deeds, chronicles, and matter requiring careful preservation being still entrusted to it. Presentation scrolls and so forth are made of parchment, and the coming of electricity for a time produced a sudden demand since parchment which had always been employed for candle-shades could be used similarly for electric light, though it did not withstand the heat of gas. The most unexpected market for English parchment was found by a small works in East Anglia. The place was small and almost medieval in its primitive simplicity, yet the parchment so made was being exported to the United States to be issued as graduation certificates to one of the newest of modern schools.

Parchment-making begins with the splitting of the dressed skins, which leaves the inner side free for chamois leather, while the glossy surfaced topside of the sheepskin is used for parchment. Unlike the extremely elastic chamois, the stiffer parchment can be made from almost any variety of sheepskin that has the requisite clear surface, though some breeds of sheep are liked better than others by individual makers. On the whole a well-fed Down sheep or mountain sheep give the best skins, but some of the bigger sheep breeds give large sheets of parchment which cannot be used because of excessive veining. Characteristically, the Lincolnshire sheep skins suffer from this defect, and also some of the Kent (Romney Marsh) sheep.

Nowadays the dressed skin reaches the parchment maker already split usually at the chamois works, so the dressing, lime preparation, and so forth have been already described in the previous section on chamois leather. When the skins arrive, already a clear lime-bleached white, clammy and pliant, they are at once washed. The parchment maker's yard has, of necessity, a stone or cobbled floor and drains to carry off the water, the only other apparatus being large wooden frames approximately 6 feet by 4 feet, very stoutly made, and pierced by squared pegs to which are connected strong lacing-cords.

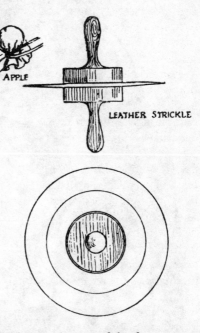

LEATHER STRICKLE

In a pile are small lumps of white skin, roughly the size of golf balls, and technically known as 'apples'. They were originally made from a small tuft of fine shavings wrapped into a scrap of damp, trimmed-off parchment. These small balls are pushed into the edge of the soft sheepskin, forming a knob or hold, around which the thongs of the frame pegs can be twisted to secure the skin to the frame (see sketch). Later, when the finished skin is cut from the frame, these small knobs, roundly compressed and dried, are cut free, and used over and over again, till they become, through successive layers of parchment, too large, when they are thrown among the 'trimmings' to be boiled down by the size-maker. The sheepskin, having been secured in the frame, is stretched uniformly to a very high tension by turning the pegs and winding up the thongs. This process calls for skilful appreciation of the elasticity of the skin by the spacing of the tension points around it. In the drawing we have shown the spacing of the pegs in one particular frame, but it varies in each works. The skin,

PARCHMENT

tightly secured, is then shrunk with boiling water, which is flung at it again and again, using no more complicated apparatus than a dipper and an old boiler. The large circular or semi-circular scraping knife is then brought into play and the skin pared down with long sweeping strokes. The

188

processes of scraping, thrashing with boiling water, and drying, are continued till the skin is a fine, clean, even parchment.

The finishing process is equally simple. It consists of brushing over the skins with a thick cream of whiting made with a solution of ordinary washing soda, and the skins are then hung to dry in a stove-heated room. Any vestige of grease dissolved by the soda is taken up and held by the whiting, which acts like blotting paper. The skins are then rubbed with pumice, washed clear of the whiting, usually given a final ' finish ' with pumice and other medium, and set to dry in a shaded draught. The drying room is usually a long lattice-sided upper shed. The light filtering through in dim bars and the wind drifting between the drying skins and rustling the scraps on the floor makes this drying room an eerie place ; in the quiet gloom, the parchments look like stored samples of family ghosts. . . .

When completely dry the parchments are cut free of the ' apples ', and packed flat under pressure. They are not usually trimmed much before selling, as most users of parchment prefer to place their cut according to the grain of the skin.

§ 4. *Sheepskins*

Sheepskin coats were common wear in medieval England and in early MSS. practically all English shepherds are shown wearing sheepskin cloaks and hoods. Probably lambskins, finer and softer, were also used when obtainable, though it would be a very special cause or accident that sacrificed a lamb, when mutton was so valuable.

A fourteenth-century writer gives a pleasant description of how to make a fine sheepskin cloak. In those days, large checks and diagonals were fashionable wear, so the needy countryman is advised to take sheepskins, black and white, shave them smooth, lay them flat, one above the other, and cut through the layers, thus shaping them into diamonds. Then the black and the white skins were to be sewed together alternately, wool side out, and the whole pressed and well lined, and the poor countryman could flaunt as fashionable a cloak as any man-about-town. Thus we see that the shaded wool of the black sheep was often used with as excellent effect in leathern wear, as in weaving (see Illus. 48).

Nowadays several chamois leather factories have a side-line in clipped wool skins to make up into gloves and winter gauntlets, and one of the largest country works for sheepskin goods is at Glastonbury where such goods have been made for centuries.

This natural depot for Cotswold skins has an excellent water supply. The works are large, and of modern design, with visiting foreign specialists, laboratory departments, and all the equipment of a town factory. Thus, together with the adjacent boot and shoe works, they are beyond the scope of this book, but the country people themselves often prepare sheepskins for their own use.

In one farm I know, on the Welsh coast, the new linoleum shines under deep, soft white sheepskins, trophies of prize-winning sheep, or household pets. They are lovely rugs, creamy white, with a silky bloom on them, and into them the bare foot sinks as into a warm snow-drift. As these skins are usually from a Show or prize-winning animal, they have been washed on the sheep, so the yolk was able to come back into the wool. This may have something to do with their extra gloss and softness, as they need less drastic scouring after skinning than unwashed pelts which cannot regain their quality when they are once removed from the sheep. This is a small point, but as it is the only suggestion the farm people could find to account for the superior quality I faithfully record it ; the same reason is assigned to the damp-resisting quality of some of the Welsh wools.

When preparing sheepskins, whether washed on the sheep beforehand or not, the country folk give the pelts a good steeping and scraping, and while damp nail them flat to a board. A shed door serves, but a perfectly flat surface is better. The subsequent treatment then varies considerably and here are some of the ingredients I have found in use for curing : lime, sheep dip, boiled bran, soft soap, weed killer, piss, ' harnessical soap ' from the chemist (I was unable to ascertain if this is an ' *arsenical* ' preparation or simply a ' *harness* and leather soap '), salt, sulphur, and the petrol (left behind by summer visitors). In fact, every maker of sheepskin rugs uses his own recipe, but the general instruction common to all methods seems to be, ' give it a scrape when passing.'

The final dressing in most cases is lime, and the final finish (to get pliability) is to beat it with a stick or to set two lads to pull it backwards and forwards across a stone wall. Since the advent of packet dyes, some enterprising workers have tried colouring the rugs, but except for the softer tints, this is not very successful. The curry comb, horsetail comb, and stiff

wire brush are needed to fluff up the wool, and after trimming the nailed edge off, the rug is considered finished, as sheepskin rugs are usually unlined.

Sometimes a sheepskin is used to ' pad ' a farm seat, or to soften the rub on some harness friction point. Once I saw a strip wrapped around the shaft pole of a double harness wagon, to prevent rubbing by the mare who ' pulled in '; and a young or inexperienced horse is occasionally given buffers of woolly skin to soften the hard edges of unaccustomed harness. So for many reasons a farmer will often lay by an odd sheepskin.

§ 5. *The Saddler*

Although the saddlers' may be a dying trade in the country, it cannot die completely out—yet. The motor may have killed horse transport and may ultimately end field work by horses, but while any animals are in work the saddler's shop will continue. The fine specialist saddler, who ' makes ' for some special hunt, the racing saddle-maker, even the saddle and trap harness-maker may lack customers, but the ordinary country maker of work harness and farm gear remains indispensable.

Few as these workers are in number, along every 20 miles or so of high road, in every small town or large village

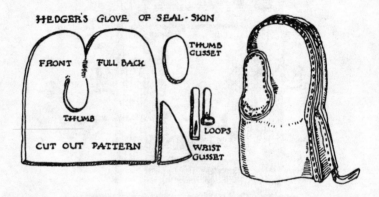

HEDGER'S GLOVE OF SEAL-SKIN

we find a saddler's shop, with its saddler usually working busily all the time. Because he is called a ' saddler ', and his work is chiefly with horse gear, the variety of things made by this countryman are sometimes forgotten by the townsman. Everything in leather, except boots and shoes, passes through his capable hands, and he gets as close to the boots as the leggings (for most of the leggings that are blocked and cut to measure are the work of the saddler).

Here is a list of things made by an ordinary saddler in a small county town : saddles and harness of all sorts, such as wheel horse tracery, canal barge harness, and all makes of plough and drag gear, including a great variety of special work-gear for hill farms ; bands for threshing and churning ; footballs, leggings, and anklets for horse and man, gloves for the hedger and ditcher; bags, braces, wrist-straps, leather cases for bottles, for field glasses, gun-cases and cartridge cases and shepherd tackle, bellows for smith and for cook, purses and straps, all manner of baggage, from the squire's solid leather travelling trunk to Mrs. Jones' little shopping bag. Also Aunt Julia's hold-all and little Willie's school satchel are of his making.

If the Vicar wants a document case for the vestry, a mother's meeting want a belt for the sewing machine, the saddler has to make them. He makes dog collars, and leads and whips: muzzles for dogs and calves (and horses), and strange

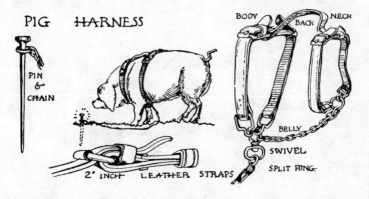

PIG HARNESS

PIN & CHAIN

BODY · BACK · NECK

BELLY

SWIVEL

SPLIT RING

2' INCH LEATHER STRAPS

veterinary appliances worked out to the instructions of the vet.

All these, and other things, he must *make*: he must also stock and be prepared to mend, webbing goods of all sorts, whistles (dog, police, and referee), horse cloths, car rugs, coloured worsted tapes for plaiting up horse tails: all the metal parts of his work, hains, horse brasses, stirrups and bits and chains: wash leathers for windows or cars, sponges for cars and horses: dandy brushes, curry combs, and dog combs, razor strops, leather goloshes for the bowling green pony, or leather harness for the roving sow: big washers for the municipal pump, or little washers for the garden syringe.

He, our saddler, stands behind all our country trades.

He mends the carpenter's basket and the blacksmith's bellows. He is, in short, working for us and with us all our days. And when we die, his work has the last word, for he has made the leather caps that muffle our passing bell.

MENDS THE CARPENTER'S PACK

Amidst so much variety of leather, it is difficult to select his most representative job, but let us analyse an ordinary

Muffle for passing bell

work-horse collar ; there are not more than fifteen varieties made in England, but we will try to describe one of the simplest.

§ 6. *A Simple Horse Collar*

It would be impossible to give examples of all the varied harness made in England. Nowadays traps and Cape carts often have the light American harness, but for field work strong horse collars are still made. These collars are not customary outside this country, so that sometimes an English-made horse collar is sent half across the world : but English collars are definitely unsuitable in hot or insect-ridden climates.

The advantage of a collar, in spite of the drawback of its weight, is that it does give a very steady draught. With band work, the varying movements and constant changing of direction of weight mean irregular tightening and slackening of the straps, with an almost slapping movement. A *well-fitting* collar tends to distribute the uneven draught and carry the weight evenly on to the shoulders of the horse.

Therefore it will be seen that the fitting of the collar is most important. A horse will do much better work and do it much more easily in a collar that fits, and great pain is caused by a wrongly adjusted collar.

It should be hardly necessary in an English book to explain

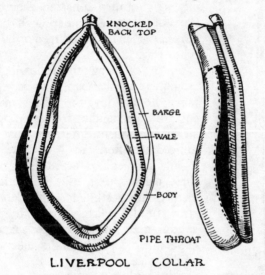

KNOCKED BACK TOP

BARGE

WALE

BODY

PIPE THROAT

LIVERPOOL COLLAR

that most horse collars are put on upside down and turned around upon the neck of the horse. But nevertheless some Irish ' pack ' collars are ' divided collars ', fastened by a strap across the hains at the top, and collars for horses doing special work such as pit ponies, fasten underneath and are completely different in design.

We can only deal here with the most ordinary type of collar. Horses in the Northern and mountain districts require collars of a closer fit, because of more varied inclines and uncertain draught. For these, too, lightness is more important than for work on the more level ground. A typical ' Liverpool collar ', in contrast to a typical ' Midland Body ', is shown in the sketch on page 195. The ' Midland Body ' is less shaped, thicker, somewhat heavier, and in every way more suited to the fat, sleek Midland work-horses in their level fields, than to the lighter and more wiry hill horses in their tilted fields (p. 199).

Even the most ' motor-minded ' person can get an idea of the great variety of horse collars if he considers all the different breeds of horses in England ; the cobby Suffolk Punch, the Devon horse, the small strong Welsh pony, the heavy, mighty Clydesdales, and many others ; and then considers their various types of work, remembering that there is a *special collar* for each breed of horse and type of work.

So in choosing the following single example it is only possible to be representative, hoping that the reader may gain enough understanding from it to appreciate all the subsequent differences that will be noticed hereafter.

The best collars are usually made in three parts. First, the wale, that is a leather tube, stuffed while damp with straw, until it is firmly shaped to form the solid frame. The subsequent shape of the collar much depends on the fine moulding of this wale.

The leather is sewn very firmly into this long tube shape, one side of the seam being left extended. This flap is called the ' barge ' and it is to this barge that the body of the collar is later secured. The straw used for filling is a fine-drawn rye

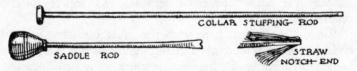

straw, which is now only grown for this purpose, for thatching, and for packing certain types of pottery. This rye straw is unthreshed, for threshing would break the hollow cylindrical straw: it is inserted gently into the wale with a long iron rod which has a very slight cleft at the end, barely sufficient to secure a wisp of the straw (see drawings page 196). The wisp of straw, about 15 inches long, is doubled and pushed up into the wale.

The packing of the straw has to be done very skilfully, first to one side and then the other of the narrow tube. Never must the alternating wisps of straw end *level* in the tube; they must overlap and interlock, each separate strand blending down into the next, and each freshly inserted piece being pressed up into, not the preceding strand, but the strand furthest from it, so that there is a continuous 'plaiting' going on inside the tube. The tube is barely 2 inches in diameter,

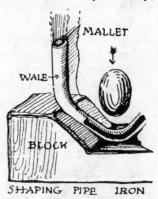

SHAPING PIPE IRON

and the whole job must be done rapidly, at one time, while the leather is still damp, so that it may shrink evenly upon the straw.

A wale might look and feel quite flawless while damp, yet any unevenness in the tight filling would show as a ' break ' or hollow upon drying out. The amount of force required is shown by the method of filling. Each strand is passed gently down into the wale and coaxed into its exact position. The whole is then inverted and the straw forced in tightly

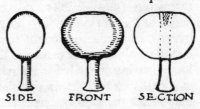

SIDE   FRONT   SECTION

by banging the iron head of the rod forcibly on to a wooden block embedded in the ground (see drawings on opposite page).

The horse's wind-pipe is set low, so to relieve pressure at the base of the throat, the wale is here shaped into a small hollow, and a thin ' pipe iron' to ensure this shape is sometimes inserted during the filling. Great care is taken to keep it embedded centrally within the rye straw, and when it is firmly embedded the correct shape is obtained by beating the entire wale with a weighted mallet on a shaping block. The least constructionally-minded person will see how evenly packed the straw must be to make this shaping process possible without dislodging the iron from its central position embedded in the straw.

The wale being completed, it is left to dry out.

The body of the collar is then made. This body is preferably also of rye straw, though in districts where this is now very difficult to obtain, wheat straw is sometimes used. The covering of the body is usually an exceedingly tough, durable woollen cloth, specially woven for saddlers and called ' body check '. Linen is also used, and I have seen very fine rush

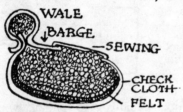

plaiting work on some old collars. Woollen cloth is soft and more porous; linen is cool and more easily scrubbed. Each material has its advantages and there are many qualities of each.

Directly under the cloth, above the straw, is a layer of ' flock '. This flock deserves a passing note, for it is one of those interesting points of ' interlock ' between changing trades. The flock is purest wool, formerly just ' refuse ' bought direct from the wool spinners, and carded by the saddlers themselves (most saddlers still keep an old pair of carding hands in the workshop, for odd jobs). Now, this wool flock is more conveniently bought from the large woollen manufacturers. It is made from the flock or fluff of the woollen looms and is sold ready for use in bolts, from which the saddler easily cuts off and shapes the piece he requires. It simplifies the job for the saddler, but it costs him more, and the small country woollen mills have lost the sale of this by-product of their looms.

The body of the collar is sewn to the barge as shown in the drawing. The cloth comes close to the wale on the inner side, and is laced to it across the straw upon the other side, which will later be covered by the facing. The stitches are single in the barge, but double in the cloth, thus shaping all to a close even finish. There must be no trace of a wrinkle on the inner surface of the body, for that presses directly against the horse's neck and would soon gall the warmed hide.

A glance at the two collars shown in the drawing will show the many points of differences between just these two, and some collars vary even more.

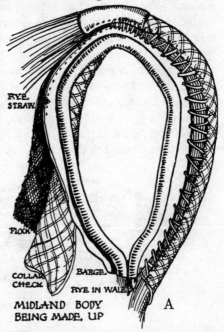

RYE STRAW

FLOCK

COLLAR CHECK

BARGE.

RYE IN WALE

MIDLAND BODY BEING MADE. UP

A

The last thing done at this stage is to knock a rope into the hollow between the wale and the barge to deepen the groove that will later contain the iron hains, and to facilitate sewing on the facing.

The leather facing is then cut and put to soak. The facing forms the most distinguishing characteristic of the entire harness. For example, a Midland facing may be small and close, barely covering the body. A Scotch facing is as wide as wings, and has whalebone and cane stiffening sewn into it, so that it sweeps in a smooth curve up to a central peak (Illus. 44).

The old Yorkshire facing had a square extension like a storm collar (which indeed it was) high behind the horse's ears.

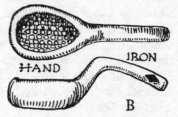

HAND

IRON

B

On to this facing were formerly put many fancy brass ornaments. The maker's stamp and trade mark are always printed on, looking forward, so that anyone standing at the horse's head could read the maker of the harness (Illus. 45).

Nowadays the Scotch collars are becoming smaller and the great Yorkshire wind-screens are shrinking away. But there are still plenty about, and the traditional differences will probably last as long as collars are made. The collar facing when made is sewn, bound, and then laid over the collar with great care. It is secured very carefully even to a straw's breadth, for the leather is exceedingly thick, and will pull slightly on drying out. On this account, where the damp leather crosses the body lacing, artist craftsmen will pad slightly with rye to ensure a fine, smooth top finish, unmarked by the criss-cross lacing below.

The facing being in exact position, it is then sewn on to the

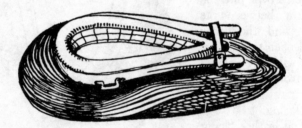

A WOODEN HAIN

barge as close to the wale as possible, the wedged rope being pulled out as the work progresses.

The saddler uses a strong, curved pack needle, and saddlers' iron, which is the equivalent of the sail-makers' palm. As shown (p. 199, Fig. B) the end is hollow to take the head of the needle, and the diamond slit grasps the triangular point and forms a firm grip to tug the needle through. This sewing of the facing requires great strength of forearm and fingers (as does all saddlery work). The sewing is done with raw hide, and each stitch is beaten and pressed home into the leather with a light mallet.

When the facing is dried out, finished, and polished, the leather work of the collar is completed. The iron hains are fitted, and their trace hooks pressed home ; the collar straps (which were incorporated in the facing when it was made) are turned neatly, and the 'simple' horse collar is made.

The iron hains are supplied by the saddler's smith (who supplies all buckles, bits, chains, and ironmongery to the trade).

In the South (Essex) ashwood hains are used : these are often painted or so darkly polished with use that I missed seeing them for years—they are fairly common in Ireland.

A brief note on harness is here added for the uninitiated.

To the collar of the ordinary cart horse are attached the trace chains to pull the cart forward. There is a saddle, or pad (it is variously named) with a deep iron groove over which the back chain passes which supports the shafts up to the right height. This back chain also takes somewhat of the draught. The cart belly-band goes under the horse to prevent the shafts rising. The breeching, or back-band, is slung at the turn of the horse's rump by bands from the back and saddle, and this back band is also secured to the shafts sufficiently closely to take the forward draught of the cart going down-hill, and keep the cart back.

SHAFT
TUG-
BLOCK

C LAM

In going downhill, the weight of the draught is taken off the collar; but if you see the collar sliding forward towards the horse's head, or if going up-hill, the breeching flapping about, then the harness must be badly adjusted or ' set '.

Carts and wagons have all manner of different harness.

Incidentally, the North favours two-wheeled carts, more than the South counties, whose more level roads take the very beautiful four-wheeled wagons. So much is this traditional usage, that the single shaft bullock wagons of the Cape are still known as Scotch carts.

Farm jobs such as ploughing, harrowing, reaping all require different gear ; but this short and very simple explanation may enable the townsman to identify the various parts of country harness when he sees them and appreciate the immense variety of work done by a country saddler even in this one branch of his trade.

§ 7.   *Ox-hide and Cow-hide*

It seems a pity that the old leathern seating of stools and benches, with plaited thongs of ox-hide, has died out of country usage. The thongs were cut from hide with the hair on, and the shading of cream and dull red, or silky black and white, against the dark oak, was beautiful.

I have found the pierced holes and slots under the present wooden seating of many old cottage and farmhouse pieces. Sometimes the position of the holes (not slots) show that the thongs were used only to uphold a cushion of padded wool or neats hair (like a Boer cartel), but I believe the wide hide strips were used as much by the English as by the Netherland workers. The chairs in the Council room at Pretoria are seated with this traditional Dutch thonging. The English design seems to use a wider thong with a plain square plait.

Hedger's gloves, and whips, leggings, and saddlery are still made from horse- and cow-hide by country folk, but most of the work is now done in town workshops.

Hair trunks are now only dusty relics in junk markets, and nothing now seems to be made in England from the decorative and durable ox-hide.

§ 8.   *The Cobbler*

The shoemakers of the countryside are few now, but still skilled craftsmen. The cobbler who ' mends ' averages two to the village, or one to the street of any small market town. But the shoe-*maker* now only plies his trade in certain select localities, usually where visitors come.

For example, in the Lakes, there are shoe-makers making mountain boots; in Scotland (though people from London bring shooting boots and ghillie wear) the old-fashioned shoe-maker still has a fair amount of custom. If you can get one of them to make a pair for you, they are worth gold untold.

You will not get the work done rapidly. He will measure you up, and then probably wait until he has booked enough orders to justify a consignment of suitable leather. These experts can afford to be very exacting about their leather, and refuse to make up material of which they disapprove. Also, the work must be left on the last for definite periods between processes.

§ 9. *Horn*

Most of the horn works of England retain their traditional positions near the terminus or by the sorting pens of the great drover routes. The huge droves of cattle crossed the hills following very definite routes to the great markets. Drovers were quite distinct from small cattlemen, who drove their own beast to the routes, for the drovers were themselves men of some substance, holding important positions in the country-side. Down the centuries while the peace of the countryside wavered and varied, the cattle drovers had to plan out a safe route for the herds, evading a marauding feudal lord, or an acquisitive district, and always following land where water could be got, and fodder could be begged, borrowed, stolen, or, of necessity, bought.

Considering that the large droves might consist of upwards of a thousand beasts, the effect of these annual migrations on the routes and local industries dependent on them can be realized. Such a route joining the Shropshire and Hereford drovers from the North, is marked over the Welsh hills, and converges on Gloucester and Bristol; other routes terminate in the Scotch Lowlands, and still others mark the way of the Irish cattle from Holyhead across Anglesey, and from Liver-pool across the Pennines.

Because the great slaughter-houses and the hide, horn, and offal works were, by law, confined to the far side of the city, and must drain into the river *below* the town, most old horn works, tanneries, and soap-boilers were formerly upon the outskirts of the towns that have since expanded to include them.

The horns of oxen and sheep consist of two parts. The outside hard substance we call horn, and the inside, flint. This flint is like a horny sponge full of fat, and fits into the outer case as a finger into a glove. The flint, before the importation of vegetable fats, was the basis of much soap-making, and after the fat for this had been extracted, went to the bone mills.

In medieval days horn was indispensable ; it was used for drinking glasses, cups, spoons, music, medicine, handles, flasks, and powder-boxes. Nowadays, it is somewhat difficult to estimate the value of horn, because there is little demand for it, yet its structure is unique, and though considerably less is used than formerly, no one yet has been able to invent a substitute for a few purposes.

Ivory and whalebone approach very nearly to horn, but re-act differently to heat, damp, or acid.

The method of preparing the horn depends upon its ultimate use. For flat purposes the tip is removed, the horn cut down one side and steamed or baked till it can be flattened under pressure. The degree to which horn may be bent or flattened is remarkable, and for combs and the like it must be so straightened. But many things may be cut from the horn in its natural form and for these it is used untreated.

Horn varies considerably in quality according to growth. For the first three years of growth a cow's horns are smooth, but as soon as an ox is full grown, or a cow calves, the ridges on their horns show age, as surely as the curves and clashes on a ram's horns show battle.

When growing, the horn is very sensitive, so that it is cruelty to strike a cow's horn sideways, though they will stand great direct pressure. The root can be injured in early growth or by accident, so that occasionally a horn will curve inwards till it reaches the skull, or droops below the jaw.[1]

Also the variety in shape of horn is extraordinary. Apart from imported foreign horns, the ordinary English cattle

[1] It is believed that the craft of transplanting horn bases so as to produce a Unicorn ' leader ' beast was practised very early by many diverse nomadic tribes.

produce a great variety. There are the long slender horns of the West Highland cattle, sturdy ones from Hereford and Shropshire, and the stubby growths of the ' shorthorns '. Welsh cattle have fine glossy black and white horns. There are also curved ram's horns, used chiefly for traditional ornamental drinking horns. (An old English dance was performed on ram's horns, the dancer wearing them as shoes and thereby performing extraordinary rocking and spinning movements.)

Rough deer-antlers are used for knife handles, and grip-pieces of all sorts. Good quality horn, not unduly flattened, will neither bend, break, nor warp. It is light and strong. This lightness and strength are its great advantages over glass, and, as it is a bad conductor of heat and will not absorb or retain flavours (as wood does) it is excellent material for spoons and scoops.

An old writer praises

> horn, as of a substauns more estimable . . .
> being neither so churlish in weight as is
> mettall, nor so froward and brittle as stone,
> (i.e. pottery), nor yet so soily in use, nor
> rough to the lips as wood is ; but lyght
> plyaunt and smooth, that with a little
> licking will allways be kept as clen as a dy.
> *Wear it not indeede that horns bee so plentie,*
> *horn-ware I believe woold be more set by than it is.*

HORN SPOONS are commonly used in the North as wooden spoons are in the South. They have advantage over wood in texture, weight, and non-conducting qualities. The best for cooking purposes are the strong, sturdy dark spoons, comparatively thick, and cut, two only, from one horn. These spoons will last a life-time, and never bend with heat, nor split with cold.

The impervious quality of horn, able to remain in the same substance for years, without absorbing flavour, or losing surface, makes it especially suitable for grocers' and chemists' scoops. ' Scoops ' are short-handled spoons. (The only wood that approximates to horn for these scoops is the hard, polished coco-nut shell, and this is sometimes used.) Horn scoops are frequently graded according to capacity—thus serving as rough measures to facilitate weighing.

For table use, the almost transparent horn is best liked,

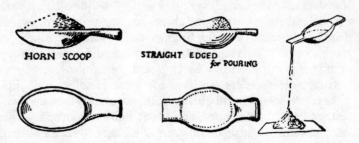

HORN SCOOP   STRAIGHT EDGED *for* POURING

and when well chosen and polished it gleams with soft moonlight gold coloured lights, amber in hue, and mother-of-pearl in its layered texture.

The slender spoons, often silver mounted, and sometimes too elaborate, that are sold at jewellers' and gift shops in the North, should not detract from the real utility of the plain horn spoon. If you have lived for any length of time in cottages where horn spoons are in general use, you will afterwards find metal spoons extraordinarily uncomfortable implements. They feel very heavy, and you will certainly burn yourself the first time you use metal

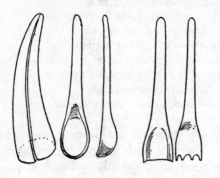

COMMON HORN SPOONS.-

after horn. The long spoon required to ' sup with the de'il ' is obviously from the largest West Highland growth of cattle. The drawings show how two spoons are cut from one horn. A ' set' of horn spoons, carefully matched, used to be a treasured possession.

Very little apparatus is used for simple horn work, a boiler and press, a fixed knife, a circular cutting saw and a lathe with various grinding and polishing buffs are all that is required—apart from the skill and knowledge of the workman.

HORN GOBLETS—

MRS. MERTON : " Does your father eat and drink out of silver ? "

SANDFORD, (Jr.) : " I don't know, madam, what you call this, but we drink at home out of long things made of horn, just such as the cows wear upon their heads."

Sandford and Merton

One of the most beautiful things I ever saw of English make was a black oak table, polished with beeswax, upon which stood twelve slender black horn tumblers. They fluted upwards, from ebony to creamy white, and there were black and white horn-handled knives to match—simple, practical, and modern, yet as old as the hills. It is a great pity that the new composition materials, from which so many various things may be fashioned, should have helped to reduce the output of the genuine horn goblet. There is even less ' apparatus ' needed to form goblets than spoons—a saw, lathe, and polishing buffs are all that is required.

BLACK WELSH HORN

Apart from the decorative value, travellers find horn light in weight and often more effective for use than metal or glass. In this connection I remember a very English incident in a small horn factory in Gloucestershire. A workman who had been in the trade from boyhood, ' and his father before him,' said that his grandfather, ' before that,' ' had a busy time *a few years ago*,' when he (the grandfather) ' got a sudden rush of orders for medicine glasses for the Crimean War '.

Because as soon as the doctors arrived abroad all the medicine glasses were found broken in the knapsacks, so then ' all army medicine glasses had to be made of horn '. It was

207

' a great rush of work in that small place in those days ! ' I remember I had spent a happy afternoon in that country workshop, and now the place was very quiet, the workers had gone home, a bee drifting in through the open door zoomed its way across and out through a broken window. On the odd tools and the worn benches, the dust of the day's work and the fluff from the polishing lathes was drifting down.

I remember how the workman stooped slowly, and hunting and rooting in an old oak bin under his bench, produced one of these old glasses for me. He wiped it, and stood it down on the bench. It was 3 inches deep, by about 2 inches in diameter, very dusty. He looked at it reminiscently, ' Very clear horn, these medicine glasses had to be—this one's darkish, expect that's why it got chucked out ; it's been there ever since the Crimea . . .'

He chucked it back into the box, together with a short shoe-horn, a snuff box that had warped, and something new that looked as if it might be a piece of aeroplane fitment. The horn, and the workman, were the same : only the wars change.

Some drinking horns have glass bottoms sprung into them. Horn has considerable spring, and the skill of the craftsman uses this property in getting an exact fit for the inserted bases.

CHAPTER SEVEN : WOOL AND FEATHERS

§ 1. *Wool*

THE COUNTRY woollen industry if treated in its entirety would fill this book. Therefore, only three of its most representative branches will be described in this chapter. They are :

(*a*) The most primitive " hand *spun* " work made in an Irish cabin.

(*b*) Plain work that is hand *woven*, from a Scotch workshop.

(*c*) Patterned Blankets from an up-to-date country factory running on water power (with an auxiliary oil engine), in a Welsh valley.

As the first deals with undyed ' natural ' cream frieze, the second with fine dyed tweeds, and the third with coloured and *patterned* blanketing, these t h r e e combined will cover most processes. There are many variations in different parts of the country, but, from the very wide range known to me personally, I select these three as being best fitted for my purpose—the description and explanation of g e n u i n e country methods.

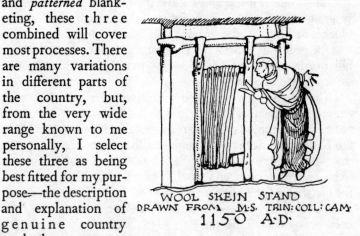

WOOL SKEIN STAND
DRAWN FROM M.S. TRIN: COLL: CAM.
1150 A.D.

Historically, the prosperity of England has been wound up in her wool. Weaving is one of the fundamental crafts, like basket-making, and as early as anything was made in England, cloth was woven. Our damp insular climate is an asset for growing fleece, and the popularity and price willingly paid for early English wool across the Channel, proves that the quality was good—as good as could be at that period—and that there was a demand for it.

Actually one of the earliest fragments of wool cloth that we possess demonstrates that the wearing of a hair shirt was a really severe religious penance. This scrap of wool cloth (believed to be twelfth- or thirteenth-century) is coarse, and full of long, goat-like hairs—though as this scrap is used as lining to a heavy bible box it may not be a fair sample of the softer weaves. In corroboration, however, the oldest breeds of our English sheep still show these strong hairs among the wool and it is a definite characteristic of certain breeds. The small Welsh sheep have a bristle of it around their necks and down their backs. The old dun-faced Scotch sheep have it too, and the coarse strong-boned Marsh sheep. Many a time old shepherd friends on the hills have shown me how to look for this 'old Adam' in certain 'families' of ewes, and watch for it again, to be seen in small definite markings in their lambs.

Old manuscript drawings show the earliest of our sheep as straight-horned (ewes were horned also), with wool of the 'lock' or long-wooled type. This long-stranded-wool type was for centuries considered the best suited for wet and snowy hill grazings, but of late it seems that Cheviots and other similar breeds do well there also.

I can almost discern some primitive, curly-wooled types among the sheep of the early Saxon manuscripts, but much must remain conjectural about those early breeds. We know the Cistercians were good shepherds and did much to improve the wool, and there is early evidence of cross-breeding to improve both mutton and wool. There is an intriguing record, for instance, of the importing of a special and expensive ram at Hull, and an early mention of the Eastern rams that may have been brought back by the Crusaders. Across the mountains of Central Wales, I once found a tiny path that ran directly between two old Abbeys that had been renowned for their wool revenues; the translation of its country name was 'Tup's Way'. All research on the early history of our sheep is a golden fleece of joyful possibilities, and extremely interesting.

The Romans had a wool factory at Winchester. William the Conqueror brought over Flemish weavers who settled first in Carlisle, and later in Pembrokeshire (where their descendants are traceable to-day—their very milk-cans are of Flemish design, see page 20).

For centuries we tried to keep our wool for our own spinners, but there was a ready market across the Channel, and extensive smuggling made it a hopeless task. In Elizabethan days, wool was exportéd, but all manner of legal enforcements and fashion-devices were tried to encourage the wearing of English wool. National coercion persuaded English people that to 'wear more wool' was healthy, lovely, and in every way desirable, and it appears to have flourished, as advertising campaigns do to-day, for in Charles II's reign, one even had to be buried in wool. Here is a legal form of that date, which had to be signed by the officiating clergy and witnesses.

" . . . . . . (*signature*) of the Parish of . . . . . . Taketh oath the Deceased was not put in, wound, or wrapt up, or Buried, in any Shirt, Shift, Sheet, or Shroud, or in anything whatever made or mingled, with Flax, Hemp, Silk, Hair, gold or silver or other *than what is made of Sheep's Wool only*.

" Nor in a coffin lined, or faced, with any cloth, stuff, or other thing whatsoever made, or mingled, with any other Material *but Sheep's Wool only*.

(*Signatures*) of Attendant . . .
and Witnesses. . . ."

In 1660 it was forbidden to export wool, and by 1774 it is reckoned there must have been more than 10,000,000 sheep in clip in England.

The *quality* of the wool is an interesting development also. We mentioned the long staple and rather hairy wool of the early centuries. Selective breeding, always eating the hairy sheep and keeping the ewes with the best wool, gradually produced sheep with wool that could be made into a very silky, rather resilient cloth, somewhat like the old-fashioned Alpaca, and the ' lustre ' was valuable when silk was so expensive. The ' lock ' breeds of Lincoln the Dales (Yorkshire, especially Wensleydale and Wharfedale), and the North and Leicester Wolds sheep were considered best in clip.

The eighteenth century saw the advent of cotton wool, and then all textile technique was gradually changed; a soft curly wool was in greater demand, and the softer curly South Down wool and Cheviots came to their own. Oxfordshire, with its ' blanket' fleece has remained fairly stable.

Nowadays, the change from ' mutton ' to ' lamb ' has had its effect on the wool trade—for lambs' wool is ' first shearing ', and the first wool has pointed ends and is different in texture, and if taken from the pelt (skin removed from a slaughtered animal), a different length.

It is believed that at a very early date the Cheviots were crossed with Spanish sheep (as usual the wreck of the Spanish Armada is held to be responsible), but there is no real evidence for this, and it is far more likely that sailors plying their trade from the South of Ireland northwards first tried the experiment. By the Middle Ages, the records are plentiful of important work in wool breeding in England, and of importations from Spain and elsewhere, and in a long wearisome medieval argument between the goose, the horse, and the sheep, the virtuous sheep is wordily victorious, in a woolly blend of pious references to the Pascal Lamb and Practical Commercial Considerations.

The earliest wool was probably *plucked* at the time of natural shedding (moulting). This is still done in the Shetland Islands. This process is called ' rooing ' and helps to produce a very fine grade of wool, and may, in itself, have done something to refine the fleece. Rooing is no more painful than shearing, as the year's crop is naturally thinned close to the skin. The women hold the sheep down, as the men do in shearing, and simply lift the wool off in handfuls. Shetland wool is certainly extremely fine, and light in quality.

The earliest hand-spun thread was made with distaff and spindle. One weaver kept many spinners busy, so that every woman and child spun continuously. It was not only an occupation for the fireside—they spun while walking about, minding their geese, their babies, or their pot-boiling; it was

212

as automatic as the knitting done to-day by the Scotch and Irish women, while walking the hills, or riding on their donkeys (or by the working girls in towns).

I have only once seen distaff and spindle used, and that was accidentally, because the spinning wheel was in use for another thread. However, as it is still occasionally employed, I have described the process later in its place.

At first, all cloth was woven on a simple loom by one weaver. Now the distance that the shuttle can be thrown to and fro by one pair of hands is a short one, as if you throw a 'catch' between hands held out before you, in a straight line you will find that you can only get a range of about 26 inches at the most. So all the earliest weaves were narrow widths, and a study of the little kilted figures skirmishing through the early manuscripts show them wearing the type of clothing that could best be made from pieces of this width.

FROM M·S·

NARROW WIDTH 11TH CENT·Y·

Later, the loom grew wider; two people worked at it, and there were new attachments for the shuttle (giving longer ' fly ') so that double-width cloth could be made, and the ' cut ' of clothes much altered.

Cloth soon became the subject of complicated legislation —certain widths and lengths were fixed, and certain weights for texture, and in order to ensure a more closely-woven cloth some cloths might only be woven by strong men (ostensibly for the good of the cloth, but involved with medieval employment problems). Later, the weaving sheds had all to be on the ground floor, so that inspection might easily be made. It would fil1 this volume if I were merely to enumerate the simplest of the wool laws at this period. It is interesting to note that the word ' fent ' used in the North for odd lengths of stuff is a relic of the legislation that ordered the ends of a roll of cloth to be left *uncut*, till after the searchers had measured it. If a roll of cloth was found deficient, this end was removed, and the cloth could not go to the ordinary market, but had to be disposed of separately. The length of the yard stick (originally an arrow) was fixed, so to-day the little dress-maker, who measures her yard from out-stretched finger-tip to ear, is using the old archer's yard—the pull of his bow-string. As the archer's own size slightly altered his

'yard' for uniform cloth, he was allowanced 3½ measures by *his own yard*. After fixing the standard yard stick, there was more trouble, for the measurers, using the stick, would mark the place with their fat thumbs, and put the stick down again on the further side, thus gaining one inch—and in time, an ell, or hand's breadth (give an inch and take an ell).

Dyestuffs were always a weakness in English cloth. Continuously down the centuries come complaints of our 'bad dyes', and, notably, during the Tudor period, come instructions to all seafarers to be on the look-out for dye-

The straight line sighting along the arrow from hand to eye was each man's own yard.

stuffs—and so make every effort to improve our dye works—'for the dye spoils the reputation of our English cloth' (*Hakluyt, Purchas*, and others). Nevertheless dye is still one of the problems to-day for which we often invite expert foreign advice.

The country people used all manner of curious ingredients to dye their wool. Some ideas crossed the Atlantic, and amid the greater possibilities of the new world, and under the tutelage of the Red Indians (who made some excellent colours), the early American women settlers perfected a fine selection of ' home ' dyes (some of which have returned to England again, none the worse for the journey).

I describe some of the dyes in the second section, so I will relate here only one incident in illustration of the variety of ingredients employed in dyeing.

I was in the west of Connemara with a German friend who was doing research work on dyes. We had hung over the peat reel in the Irish shed ; we had watched the pungent brew, boiling up through holes in the wooden board, till the process was apparently complete, and had just gone to our tea, when there came a fearful crash and clatter from the shed. We fled out, but the German worker got there first, and peered in. . . . He came back, the steam dimming his glasses, and the perspiration of perplexity dewing his brow. ' She haf now put in many mussel shells ! ' Apparently they use the big river mussel shells, among other odds and ends, as a mordant for some colours.

The Western Highlands and Shetlands still use the natural coloured wools from the black sheep, which are every warm shade of sepia. The various types of wool are well known by the country people who spin and weave them ; for example, the long, rather ' wiry ' wool is used for hard wear in worsted stockings, and the soft lamb's fleece for baby-wear.

The women of the wet West knit up special socks of undressed wool which is thick with natural yolk (waterproof, and somewhat oily) for their fishermen, and stockings, too, to wear inside their sea-boots. These hand-knit fishermen's socks by generations of the West of Ireland women have lately been discovered by sports enthusiasts to be ideal ski-ing socks. You pay 9d. for them ' off the pins ' at the cabin door, and 12s. 6d. in London.

The very greasy wool from around the udders of the ewes is also used for special purposes. Since medieval times, this udder wool has rendered a natural lanoline dressing. The medieval doctors extracted the lanoline, but shepherds' wives to-day use the impregnated wool. It is good for bruises and chafing ; when washed and dried it is like a medicated cotton-wool, but creamy in colour, and lighter in texture than cotton. It makes excellent padding for splints.

215

§ 2. *Carding and Spinning Wool*

O leese[1] me on my spinning wheel,
O leese me on my rock and reel ;
Frae top to toe, that cleeds me bien
And haps me fiel and warm at e'en !

BESS AND HER WHEEL   Burns

Textile work is strongly rooted in tradition, perhaps
because it is largely a woman's job, and women stayed at
home.

Spinning-wheels can be found in most of the Western
Isles and North Britain, but turning over the pictures in my
memory one particular spinster's cottage stands out. I have
by me now a piece of the cloth, and the scene before my eyes
is as real as its texture between my fingers. The place was in
a ' bog ', that is, a rough level plain filling a trough between
mountains. The prevailing west wind spilled its burden of
water continuously over the land, and one would suppose
that the sheep breeding on those hills had grown waterproof
in self-defence.

On the day I describe, the long light afternoon sun was
falling west, the rough bronze heath was gashed with dark
rifts of heather and bright green swampy places, and there
were small rounded sunlit pools in the peat, so that the bog
was flung down level below the hills like a battered bronze
shield studded with gold.

A mile from the road the track crossed a tidal ford ; some
rocks made stepping-stones at low water, but the steep banks
on either side were worn into slides, where the donkeys had
gone up and down, descending, apparently, ' down sitting,'
judging by the state of the bank, and the behind of the donkey
we overtook. She was a grey donkey, two panniers of wool
hung either side, and a small boy drowsed between them.
The donkey's unshod hoofs had splayed forwards like a cow's
and made a padding sound on the soft track. Small translucent
white pebbles broke clean and white from the crumbling
black peat. Quivering on its fine grass stems some bog cotton
caught the light, almost luminous. The peasants gather bog

---

[1] Term of congratulatory endearment.

cotton (' eriophorum ') to stuff their pillows, but the staple is too short for spinning. Slowly and sleepily we followed a bog path, built up of ' rock ' and packed peats, to a sloping island of turf-covered outcrop, on which stood the spinner's home. The long, low cabin was washed a dull rusty pink, the thatch was dark, and the house-leek, planted along the lines of the thatch, had rusty red rims around its watery lobes. On top of the high Irish bank beyond grew willows, yellow stemmed, with slender silver and gold leaves throwing lavender shadows over the pink-washed wall. Water, gold and green in the mosses, talked at the foot of the bank and ran past the half-door of the house. The door opened on to a stamped earth floor and someone had put down an armful of cut rushes across the entrance.

The sound of the sea, far away, filled the air ; the light was soft as milk, and wet with misting rain, and through the open half-door came the drone and whirr of a spinning-wheel (Huck Finn affirms that a spinning-wheel makes the lonesomest sound on earth).

Within, the cabin was divided into two parts by the stone chimney and open fireplace with its dusty black hearth and hanging cranes and pots. The further half was the bedroom, and a second sleeping loft was reached up an open ladder from the living room. The loft was only a plank flooring under the rafters, set across the cabin walls. The end was open, with a curtain drawn across, so that people walking there looked as if they trod a very high stage, floodlit by the low firelight. A bed was made up beyond the fireplace ; there was a pieced calico cover, sheets of thick linen, and on a folded blanket slept a baby, cushioned against a sleek, maternal-looking tabby cat. Beyond the bed foot was a chest of drawers, of rough wood and clumsily built, but painted scarlet and white in chevron patterns, with an elaborate white toilet cover over it. It had a small made-up altar on top, with a red Sacred Heart lamp, burning. Bright oleograph portraits of the entire Holy Family were on the walls, with palm fronds and a china holy water stoup. The holy water plays its part in the spinning process ; if the wool is difficult to spin they sprinkle it with oil *or* holy water. There was some kept for the purpose in a medicine bottle, with a china cork sprinkler.

An inflated hen rose with indignation as I entered, and swept her brood under her beneath the bed, crouching there, her red poll just lifting the vallance, all the time I was there. Two windows, deep set, were built either side of the open door, and the sun laid pale squares over the brown earthen floor.

The wheel faced the end window, and before it a gaunt spinster was stepping back and forth. Her bare feet were

THE WHEEL BY THE WINDOW

soundless on the earth floor, and the grey strands of her hair waved to and fro as if they, too, were being spun. On a bench by the other window an older woman sat knitting, spinning with a bobbin at the same time. She was twisting up just enough wool to finish the toe of a thick grey sock, and so working direct from the spindle to the needles. Her man sat on the window seat, minus one sock as she worked. The two women wore woollen skirts of the home-spun cloth, cut very full and gathered clumsily on to bands. The man's shirt was of the same cream coloured cloth; it looked comfortable, with wide arm-holes, and narrow cuff-bands and neck faced with smooth linen. It fastened with large, single horn buttons, and had two big pockets. His trousers were the same weave as the women's skirts, but of white wool, and brown warp, so that they looked speckled. The trousers were of hairy and rough cloth but worn smooth as an egg behind, for rowing a curragh puts a fine polish on you.

There was a solid table and some chairs. An iron pot, full of rain water, over the hearth was heating for the child's bath and the man's wash.

In spite of the earth floor, all the house was very clean : the bed in which I slept was soft and warm, the pallet bed in the loft had plenty of the same thick cream blankets, and the bed by the fire was even more warmly covered. There were no books, for none of the three could read : the priest did their sums for them.

The two women were the spinners, and made the thread ; the thread was then carried to the weaver across the bog, who wove it on his loom.

Many spinners were needed to keep the one loom going. They paid the weaver in kind, and in extra thread. It was a complicated communal barter system, for the entire bog ' spun ' around that one weaver, so that he waxed exceeding rich. The entire community bought back and wore the cloth he wove, and drew revenue from what he sold away.

The whole craft of spinning varies very much, but I describe the process exactly as I saw it done here.

First of all the wool was cleaned and sorted. The fleece from slaughtered sheep being longest, is kept separate from sheared wool. The former makes the strongest worsted and so is chiefly used for stockings. Ordinary sheared wool is divided into long and short ; fine wool and lambs' wool are kept separately, but nothing like the complicated commercial grading is attempted (Illus. 47).

Wool-combing was once an independent craft, employing many men. The great red oil jars that serve as signs for the London oil shops (and the very name ' oil shop ', too) originate from the wool-combers. They used one pot to three or four ' benches ' of workers, and a charcoal fire inside the pot heated the wool combs. The oil was used to sprinkle the wool in working, and the hanks were 'drawn' over the heated combs, which were raised arm-high on the walls of the workshop.

Wool, by reason of its nature, 'binds' naturally. The texture is most easily likened to the tiny hooks that keep intact the elastic plume of a bird's feather—once this is broken, you cannot bind it together again. So with wool: if combed as it grows it can be drawn out into long smooth strands or slivers and then spun, but once the wool fibres

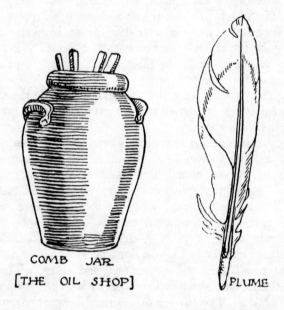

COMB JAR
[THE OIL SHOP]

PLUME

are broken or bent or tangled their quality is destroyed. Thus the 'combings' left on the wool combs ('noyles') are plucked off and put into another bin to be used for stuffing upholsterer's work, or saddlery.

Dressed slivers of wool are smooth as silk, and several feet in length, each hair lying evenly in one direction, and all wool is thus prepared before spinning is begun.

These spinners had to prepare their own wool, and after a preliminary drawing in the shed, they finished the process in the house with carding hands, and the carded wool was laid in a box ready for use. In the winter, when it grew dark early, they carded when they could no longer see to spin.

They washed the wool in a clear stream, treading it with

CARDING HANDS

their feet, and rinsing on wooden boards. They wrung it out by folding it into a thick twist, looped over two sticks, hanging it to dry on a fine day in hammocks of old fish net, rigged up in the wind.

WOOL WRING

When I first entered the cabin on this particular day, the spinner, doing a long spin, was using the drawn wool direct from the box into which she had carded it, and spinning as fast as she could wind the thread on to the spindle.

As the process at the wheel is a development of handspinning with distaff and spindle, we will describe that first. A length of the carded wool is pulled down from the distaff, and the further end secured to the spindle ; it is then *spun* swiftly, with a dexterous twist, till there is a close, congested twist compressed between the spindle and the thumb and finger holding the pulled-out wool. Exactly at the top of this twist—at the moment when the spindle would, ordinarily, begin to reverse and spin back—the spinner drops the spindle or stops it against her skirt or foot ; at the same instant, she releases her finger and thumb, so that the twist runs up the spare wool she has teased out, and another length of thread is ready for winding on to the spindle on which it is secured by a notch, or slip loop, and the process is repeated.

A good spinner takes care to overlap her twist so that the thread is evenly strong, and skill is needed to get an even sized thread.

In East Anglia the Flemish spinners had their own method. They called the distaff a Rock, and they spun the spindle and wound the thread on it with a quick rolling movement against their knee, or sometimes they had a little smooth leather strip (called a trip skin) hanging from the waist to roll the spindle against their thigh. The Flemings were, of course, renowned

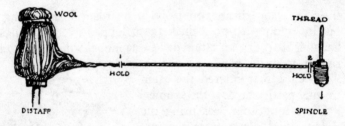

for their spinning and weaving, and could do a wonderfully fine thread.

On the wheel, this process is exactly the same, but done more quickly and continuously. The wheel turns the spindle, on to the ' pole ' of which the reels are pushed, and where the twisted thread is wound. The later development in wheels that automatically twisted the thread as it reeled it up was used in factories, but, I believe, never by country women.

The spinster moved back, as she made the spindle twist the thread—sideways and forwards, as she let the spindle wind it in. The stroke of her left-hand that kept the wheel turning fell naturally between these two actions.

The long spinning galleries in old houses were used when the spinner used to walk backward from the wheel; this gave a long drawn-out thread, but I have never myself seen the method used (except for rope work).

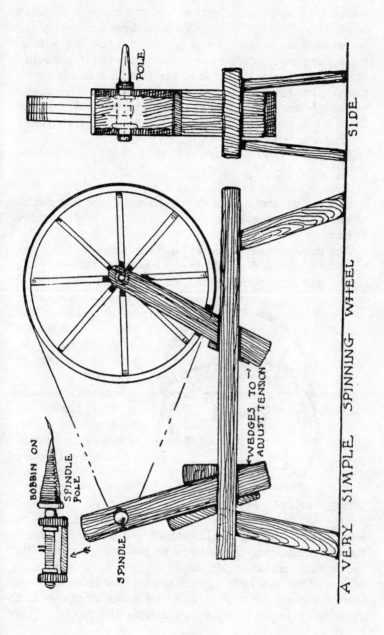

POLE

SIDE

A VERY SIMPLE SPINNING WHEEL

WEDGES TO ADJUST TENSION

SPINDLE

BOBBIN ON SPINDLE POLE

The long staple wool that is best for knitting the thick fishermen's stockings and socks is sometimes used unscoured —that is, with the oil used in preparing it together with as much natural yolk as there is in the wool, still incorporated in it. This makes a rather ' dirty ' knit, but is certainly waterproof. The ' prepared ' home-spun of commerce is frequently so much better cleaned, and so much more smoothly finished, that if it gets really wet, it swells up like a sponge and becomes too heavy to wear.

§ 3. *Dyeing and Weaving Wool*

In the first textile section we described the home-spun thread and cloth because it was woven and worn by the entire district.

The weavers of the district I visited were more sophisticated, since they could and did obtain easily ordinary machine-made cloth ; but they preferred the local home-woven cloth for their own hardwearing clothes, and any cloth that was over sold for a good price to the autumn visitors.

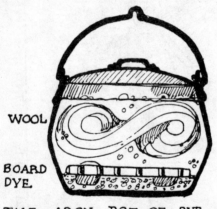

WOOL

BOARD
DYE

THE IRON POT OF DYE

The work was done in a Highland Glen of great beauty, but it was not a ' show ' place—and the weaving shed was inconspicuous and entirely utilitarian. It was of grey stone, set well away from the small cluster of houses gathered around the bridge and inn. A stream went past, which worked the fulling hammers on a water-wheel. There were three looms set up. One was standing, because the weaver was away ; the second was set with a black and white straight weave, on which a boy-apprentice was learning the process. The third loom was being worked by the owner of the premises, with a very pleasant greyish tweed.

There is very little need to describe a loom at work as the process is so well known. The warp threads are set up the length of the cloth, and the treadles going up and down lift these threads alternately or in a definite sequence, imprisoning the woof threads which the shuttle, flying to and fro, carries between. The shuttle is fed with the reels filled either by the spinner, or by the winding machine, and a frame, pushed back against the threads, packs them tightly into a firm cloth.

There are various makes of looms, and contrivances whereby the finished cloth is automatically wound up and fresh warp released, but all these are technical details.

We went next to see the dyeing expert, who lived 10 miles off. Much of the characteristic smell of peat smoke and burn-water in Harris and other home-spun tweeds comes from the crotal used in dyeing them. This grey lichen is gathered off the heather and rocks,

DYE BOARD

and has a strong, somewhat hempen smell, that returns when the tweed is damp, and never quite goes off.

The dyestuffs used by the country people long ago were extraordinarily varied, but few ingredients now continue in use, and the difficulty for researchers lies in the fact that the best dyers are usually least capable of explaining what they do use or how they use it. Of dyestuffs I have *seen* in actual use, I remember seaweed, soot, shells, chamber lye, black-berries (berries and leaf-tops), rowan, and black-currants, onion skins (yellow), elder bark, crotal, and half a dozen or so bog plants. Some things are dyed for the colour—others (like mussel-shells) are mordants.

The method of dyeing wool is worth describing shortly, for it is very ancient; I found the same method described exactly in a German household book of the fifteenth century. The wool is washed and damped evenly. The dyestuff is

225

bruised and laid at the bottom of an iron pot, over which a wooden board, pierced with holes, is laid and weighted down with stones. The wool to be dyed is kept stirring above the board, and the dye bubbles up and over it.

In the case of fruit stain and other dyes, salt water is used to set the colour. The day we were there, onion skins mixed with shredded peat were being used to obtain the required shade of dull orange.

The mixture took all the afternoon to boil, and the wool was afterwards laid down under stones in the burn to rinse. It would be fetched up late in the evening, and then hung to dry over the peats for the night.

In contrast to this simple spinning and weaving, with its primitive barter, or direct sale by the maker, some of the country woollen mills are most advanced and up-to-date concerns, with travellers and agents and show windows.

Some of the Welsh mills make excellent blankets of coloured grey and brown patterns, with touches of scarlet or green, and where the traditional squared designs and natural colourings are used, in these blankets are as excellent specimens as any in Britain. They wear for fifty years or more, and only improve with washing. Unfortunately these blankets have been exploited by commercial imitators whose badly designed circular patterns (so inappropriate to the straight, squared convention of the weaver) and crude tawdry colours of poor dyes, bring disrepute to an antique and admirable craft. Welsh flannel survives in small country mills on pure merit. It is unshrinkable, and is soft and extremely warm, and what it loses by comparison with some of the fine commercial 'finishes', it gains in wearing qualities. Many of the older mills use teazles for dressing, and have some difficulty in obtaining them, as teazles are not widely grown, and the crop cannot be repeated for seven years. Some Welsh mills still turn out the coloured striped clothes-flannel, worn by the older generation for their working skirts, and there is specially strong, thick check blanket-cloth woven for lining horse collars (see p. 198).

A description of one country woollen mill that is still

flourishing may demonstrate that such mills can be commercial propositions.

It is a Welsh mill. In the main building there are three or four looms which are worked by an overshot water-wheel, and two engine-driven looms are in a separate shed. There is a dye house and a fulling cellar. A large grass meadow is set around with stretching poles, and the mill stream is run through several wash bins.

The owner has two distinct markets, a steady small local trade for coarse country stuff, and a short season trade to summer visitors. The water power is low in the summer, but full in winter. So the whole mill is a good example of that seasonal adaptability that is essential to country work.

The first time I went, at high noon in the hot summer of 1933, the mill was almost empty because the entire staff, taking advantage of the low water, were clearing out the mill dam and repairing the sluice, and with crow-bars, picks, and shovels were demonstrating the adaptability expected of the country worker. They were weavers, but they were also hydraulic engineers that day.

Another windy, blowing day, the entire staff seemed to be in the 'tent' field, or tenters' yard, stretching the cloth on the hooks.

The stretching frames are extremely simple, two wooden rails set with small sharp hooks on to which the selvedges of the cloth were spiked. When all was smooth, a crank wheel lowered the bottom rail, and wind and rain did the rest. This day, on leaving the tenter's yard, I ran into a bright green boy. The dye hand had also seized the fine day for his job and his fire boy (whose job it was to stoke the boiler) had sweated for it. Apparently in a moment of crisis both workers had wiped their sweat with one of the green swabs, and (it expresses their absorption in their job) neither master nor apprentice had noticed the startling effect.

The 'dye shed' was an old stone shed with three large copper boilers, a number of tubs, and a hose pipe; the department also owned several barrows (so vilely coloured and dyed, by intent 'that no one dare pinch them off us')

227

and about a dozen spring ' beds ', that is, raised wire netting frames on which to drain and dry the wool (Illus. 46).

Washing troughs were fixed under the over-shot wheel and in the mill race. This firm used commercial packet dyes, chiefly German.

Such was the varied summer work.

Another day in winter I visited the mill and the yards outside were deserted. The sheds closed, only the wheel was turning, and I could hear the steady monotonous clack, clack of the entire mill working away on the weaving. The power looms were being used for fine cloth, and three water-power looms for thicker tweed. There was also a single blanket loom.

As a lucrative development, this mill had started a knitting department. They had several circular stocking machines, and a couple of flat knitters. The women knitters (the loom weavers were all men) were busily making sports hose, sweaters, cardigans, and gloves (partly by hand knitting) of the same dyed wools to match the tweeds made on the looms. Practically every customer buying a length of tweed bought two or more matching hose or a jersey or sweater, so that the knitting department was growing, and several machines were ' out ' in the village.

Besides the seasonal trade in tweeds and knitted goods, they also kept going a small but steady line of coarse woollen goods made for local country use. The country farmer is very conservative, and will pay a good price for exactly what he wants (judging his value by use, rather than appearance).

So the mill also sold cream blanket-cloth pants, with strong linen tape ties, and shirts of excellent Welsh flannel, made in the old-fashioned style, with an *open* square gusset under the arm. A good pattern for ploughmen and labourers, as the sweat is let out, and the cloth never becomes bulky and thickened, rubbing under the armpit. For the same market, they also made thick undyed woollen socks and stockings for wearing inside heavy field boots.

This ability to meet two distinct markets seems to be the keynote of most successful country mills. It is useless for the small country weaver to try and produce quick, cheap goods for a town fashion that changes too quickly for him to touch it, yet it is impossible to keep going only with old-fashioned country stuff. The most successful country weaver is he who specializes on fine individuality and superlative quality (Illus. 48).

And the buyer of such goods, whether he be the local farmer wanting all-wool spliced pants, or the touring motorist needing a spare rug, will like simple, good quality work, and infinitely prefer simple traditional designs.[1]

§ 4. *Knitting Wool*

The Scotch are always considered the best knitters, and fifty years ago, a fine, well-knit pair of Highland hose were worth 30s. to 40s. The big revival of knitting and crochet that began during the War still continues, so that few places

APPLEDORE JERSEY

TEN STEEL NEEDLES

now retain the old designs, but the Fair Isle and Shetland knitting and Irish crochet are still strongly traditional.

As representative of the many lesser known specialities I describe an Appledore jersey.

Appledore jerseys date with Monmouth caps in a creditable ancestry from the days of Henry VIII and are as different from other sweaters as Audit Ale differs from ordinary ale. An average specimen of Appledore jersey is worth at least

[1] Since this was written, the mill has changed hands.

£2 2s. or more if it is an exceptionally good example. A fair commercial imitation can be obtained in Bideford, but the expert would no more look at a ready-made Appledore than a high-class London tailor would keep stock sizes. Each jersey is made for a particular individual.

Usually, for the sailors, a fine, somewhat harsh yarn is used of fast indigo dye. They are knitted on fine steel stocking  needles, as many as sixteen needles to the round of 400 stitches (or up to 600 for a large size). They are knitted in a complete circle, like a stocking. There are no seams, and the gussets under the shoulder, sleeve shaping, and neck pieces are all pieced out in one pattern, a most intricate and skilful piece of work. In the whole jersey there is no join, no stitch put in anywhere, and nothing is used but the skill of the fine needles. In the best work, the tension is comparatively tight. The breadth of the shoulder is reinforced with moss stitch, and the spreading elevation for the neck spliced with special stitches.

The name of the ship and harbour was often knitted in different coloured wool across the chest. This might mean the same kind of jerseys were issued for all the ship's company. It ensured that if the wearer was found drunk he was returned to his right ship, or if he was found drowned he was buried in the right grave.

I regret that the sea-coast industries have had, for reasons of space, to be omitted (but see note on page 245), as sailors and the old seamen are wonderfully interesting craftsmen, making many and varied things. Not only do they knit, sew, tat, carve, work in rope and shell, but most of them are specially good artists in their crafts.

§ 5. *A Wool Pack*

The Chancellor sits on the Woolsack in Westminster. This is how a woolsack or pack is made in a country barn.

The elasticity of wool makes it unsuitable for transport in loose fleece rolls, so that it has to be compressed.

A woolsack, empty, is like a mattress cover 4 feet square. When standard sizes were legally in force, *probably* the northern packs were more flat than the southern, due to the narrower width of the smaller moorland fleece. The empty woolsack is hung upright from the top beam of the barn, and two packers clamber up and drop down inside. As the rolled fleeces are flung up into the sack, the packers walk backwards and forwards, inside, treading them down. As more fleeces are thrown in, the packers are raised up till their

PACKING WOOL

heads appear over the top edge of the sack. Presently this top edge barely reaches their knees, and they have to hold on to the beam to keep their balance on the upright swaying wall of wool under their feet. When full, the top edges of the sack are quickly caught across, with iron prongs, before the elastic wool swells and makes sewing impossible. Though there are iron clips made for the purpose, the old three-pronged kitchen fork was a useful substitute, as the discovery of old dinner forks jabbed upright into the topside of a barn rafter 14 feet overhead conclusively prove. I remember finding such a fork when crawling along a barn rafter to see

231

an owl. Twenty years later I checked the observation in Stephens' *Book of the Farm*.

Stephens also mentions the round stones used to secure

TWO PRONGED FORK

the rope to the 'pack sheet'. These are the common device for holding farmhouse cloth of every kind, from stack sheets to jelly bags. They are called 'apples' by the skin dressers (see p. 187).

§ 6. *Wool Rugs*

In America they collect hooked rugs. In England we use rag carpets. These have been made in England variously for centuries, often fashioned in old cottages, from the contents of the household rag-bag, sliced up and 'pegged' or 'hooked' through sacking. In some county towns, factory-made rag rugs can be bought.

The origin of this humble craft is probably northern, and its necessity arose from the cold stone floors of the North. Its development came through the 'fent' sellers, who sold off the cuts and left-over pieces from the weaving mills. Its modern continuance is through enterprising woollen merchants, who sell good quality specially cut wools of brave new dye, and advertise their use through Women's Institutes and fancy-work stores. So nowadays, the old rag rugs, which rooted over the stone flags of the cottage and grew beautiful in the warmth of the afternoon kitchen fire, may appear in modernistic design upon the parquet floors of town mansions.

The methods of making vary as much as the rugs. A long time ago they used the wool rejected by the carding hands, rough locks, and fleeces broken in the shearing, but these rugs do not exist to-day, when any wool used is carefully prepared and wool is considered the better class rug ; the rag rug having become the humbler article.

232

The pieces may be cut roughly to length before being hooked or pegged to the sacking, or they may be cut afterwards. Sometimes the worker likes to work with the rug stretched on a frame, sometimes it is held over the knee, it depends on the worker.

When 'hooked', the hook is thrust through and the loop of wool pulled up, and hooked over the two cut ends with an action like crochet. Where 'pegged', the centre of the cloth piece is thrust through, and the two fingers of the other hand are used to open the loop thus made, and pull the two ends through it. In both cases it makes a tie, or loop on the under side of the hessian, as shown in the drawings.

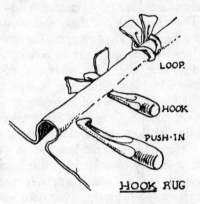

HOOK RUG

Of course, the fine new woollen carpets are very carefully cut or finished to a smooth surface, and the coarser wool or rag rugs are often sheared or clipped evenly when finished, but others made of fairly wide pieces of cloth are left rough. In the North some very fine work of this sort was used for bed coverings (something like the tufted quilts that went to America), but these are not made now.

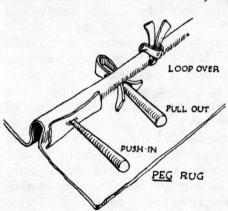

PEG RUG

Sailors often make excellent rag carpets in odd times ashore. I came across half a dozen pegging away, sitting at the back

233

of an eating house in Lowestoft. Men's rugs are usually very close set and thick. Sometimes a woman makes a rug comparatively thin, as being lighter to shake. The condemnation that these rugs 'catch the dust' is just why they are used. They *do* catch the dust, and so stop it flying about all over the kitchen. If you notice, the first thing any country woman does each morning is to roll up her hearth-rug, lift up the door-mat, and give both a shake and fling them over the fence to blow while she sweeps the floor : with open fires the hearth-rug is indispensable.

A good rag rug lasts many years ; when it is faded it goes into the dog kennel, and when the dog dies, it is burnt.

Here is how a rag hearth-rug, very well known to me (for it lay in my own kitchen) was made by a friend. I give it as typical of a countrywoman's method of working. The description is partly in her own words.

First Mrs. Roberts came over and measured the spot and then went through the rag bag to see what she could find that would 'do'. She took some large 'things', and a bundle of oddments, pushed all the cloth into an old pillow-case, and with two shillings to buy a piece of hessian, went off.

(Hessian, or sacking, can be bought by the yard in all small country shops ; on farms, they usually have an odd sack that can be used.)

Next day, Mrs. Roberts cut up the pieces of cloth and the 'things', chopping off buttons, and throwing away old linings and seams, and laid all the pieces out flat on top of each other, so she could 'see how much she'd got'. She judged she had sufficient for the entire job. Then she bundled all back into the pillow-case.

Next market day, she bought a piece of hessian (and because it was wide width, there was enough cut off to make a work apron). The 'piece' she laid flat out in the wash-house and turned it in all round. Then she set out to mark the pattern. In the design she used a burnt stick to draw with, soup plate (one), pie dishes (two), and the boiler stick to rule the straight lines.

When she had got the pattern to her mind, she 'set the black' with the smallest rub of grease. Then she 'thought a bit '. . . . Then she rolled all up, and that was all she did that day.

The next night that she had free she turned on the wireless, and started to slice up the cloth pieces. She cut two or three thicknesses at once, and made bits about 6 inches long. Because she did each piece of cloth separately, the colours kept more or less together and when she had done as much as she wanted (or ' couldn't stick the wireless any longer '), she put all the bits into the pillow-case, and went to bed.

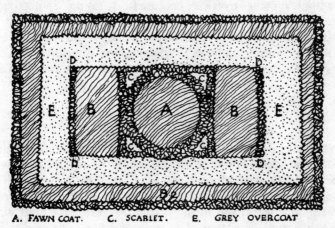

A. FAWN COAT.   C. SCARLET.   E. GREY OVERCOAT
B. FAWN SKIRT.   D. BLUE.   B.2 FAWN BORDER. Remains of A.
CENTRE & EDGE BORDER, FROM BLACK TROUSERS & SLEEVES..

This went on, odd times, for a week, when she reckoned she'd got ' enough to set to '. She worked across her knee and hooked.

The sketch shows the pattern. The middle ' A ' is fawn, the coat part of an old coat and skirt (voluminous 1910 model). The skirt made both B and B. Next she supplied a circle of black that ' had been me husband's best trousers, only they were so tight that he only used them for funerals '. To counteract any funereal effect, CCC were scarlet. (This scarlet was ' very choice ', for we hadn't much. In hunting districts old pink coats always fetch a good price at country rummage sales.) Next, because ' blue goes nicely

235

with brown ', D and D were Mary's outgrown blue gym tunic. Then came a wide grey strip, E, of ' had-been ' overcoat. After that she went on, hooking round and round, ' till the coat was used up.' She put a border of the fawn, ' that was the sleeves and pocket, and bits over, from the fawn, that the coat and skirt, that's the middle of it, was.' Then came a finish, made of the pre-obesity, ex-funeral trousers, ' because black looks nice on red tiles.' She did not line it, because I wanted it light, to shake. She allowed for it a few inches oversize, because it ' takes up ' in making ; and it fitted very nicely. That is a common rag rug.

Traditionally this type of carpet work is believed to come from the North, with the Norsemen, and remnants of such coverings, in wool, have been reported from old Viking graves.

§ 7. *Quilts*

There has been a strong revival of this old craft in the Northern border counties and in South Wales. It had never quite died out, but during the War years the country girls about to be wed, who would have made quilts, spent their time in war work, and in the period following, the craft was taken up as an employment scheme, and has since tended to pass into the category of decorative needlework.

The sturdy quilting of tradition was very different from most modern work, which is often etherealized into drawing-room chintzes and comparatively thin *coverlets*.

The finest old heirlooms were always exquisitely worked, and adequately valued. But quite excellent quilts were made and used in both Welsh and Northern districts as commonly as blankets. While I was camping near St. David's Head, a local farmer's wife brought across the fields to me a wonderful quilt, intricately worked in a pattern of ferns and squares. When I remonstrated with her for lending me such a treasure to roll on the ground, she said it was better than a blanket, because it washed better, and the damp would not harm it.

The tradition of quilt-making is very interesting. The fact that we find quilts in the spinning centres demonstrates their early making from the 'noyles' or 'card wool', that is (under its various local names) the soft loose wools left on the carding hands or teazles, good light wool of fine quality but unsuited for spinning. Much of this flock used to be purchased by the saddlers (see p. 198), who now buy the equivalent, in roll lengths from the factories.

We may trace the Flemish weavers' influence in all the quilt districts. The 'French' box beds were built into some old cottages (often under the stairway that leads from living room to loft), and the quilts were made on the principle of the huge Continental 'top-beds'. I have slept under one of them, years ago, on a Yorkshire canal barge, North Bingley

THE BARGEE'S BEDDING

Canal. (The deep-sided box beds were so coffin-like and the covers so thick, you needed a periscope for breathing.) But the single thick cover is far more practical for every type of box bed.

These wool beds were the landsman's equivalent for the medieval sailor's feather bed. (Probably because of their buoyancy, lightness, and compressibility, feather beds were provided in the small sailing ships as early as the thirteenth century—possibly earlier.)

The early woollen quilt was probably made on the principle of the woolsack. At a later date the work would be done horizontally, much as it is done to-day, but the wool remained as a loose filling, and this tradition was followed in the process of cotton-filling.

Another type of quilt was made by country people for their own use from woven blanketing. These quilts are the direct ancestors of the elaborate coverlet quilts now made to adorn the four-posters in the bedrooms of large country houses. These thinner quilts were decorative but were also extremely warm, and most early cover quilts were of blanket on the under side, and of linen on top. This was for four good reasons:

(1) The blanketing, yellowish, rough and warm, from the wool home grown and home spun, was more plentiful in

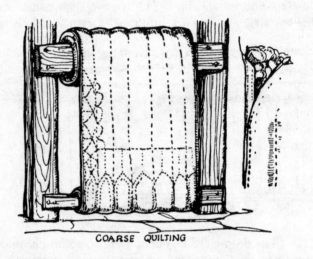

COARSE QUILTING

the cold mountain districts than linen flax. (2) For the top covering, a linen sheet could be used that was becoming too thin for hard wear, as it would get comparatively little strain sewn down on the blanket. (3) The slight shrinkage of the woollen under side in washing and drying the quilt would increase the quilted effect at each washing, and since quilting cannot be ironed easily, this is a practical point not fully appreciated by modern revivalists who use elaborate silk materials. (4) The warmth engendered by the woollen under-blanket and the non-conducting, wind-proof linen cover makes the quilt far warmer and more comfortable than any linen *encased* cover.

238

Nearly always the country things that survive longest are devised with sound common sense, and a quilt in its original simplicity was designed to give maximum warmth with minimum weight.

The drawing shows the making of an old thick quilt in a frame, but they were sometimes made on the floor of a barn, or attic, laid out flat.

An inferior kind of quilt is made, not from new wool, but from old blankets or from any woollen cloth that can be cut and spread between two calico coverings. The cloth is not cut up, but simply spread out in two or three thicknesses, one above the other. These quilts are warm but heavy.

All quilts do not rely for their pattern entirely on the intricate traditional stitching. I have a large one, purchased about fifty years ago in Skipton, a sheep town in Yorkshire, the cover of which is made up of red and white stripes, each stripe 4 inches wide, and the shell pattern being worked in red on the white stripes and white on the red. It has a wide red and white frill to hang down over the sides of the bed, so that it does not easily slip off.

§ 8. *Country Bedding and Upholstery*

There was considerably more upholstery in early English and Medieval times than is commonly supposed. The stone window seats shown in early manuscripts are very comfortably upholstered: hard wooden chairs and benches had very plump looking cushions, and good bedding was downy and comfortable.

It may be interesting to describe some of the filling materials legally and illegally used for such bedding and upholstery, as the same stuffs could be found in use in country districts right up to the present time, though the imported kapock is now taking their place.

Wool was, of course, used; not the best wool, capable of being spun, but the discarded wool, literally 'dis'carded' by the carding hands. Flock wool, either the matted, unusable locks off the fleece, cleaned and dressed, or other 'matt' wool from the spinners' and weavers' workshops might be used, or even the wool left on brambles in the grazing ground.

Then there were feathers which were comparatively plentiful : not only from hens but from wild fowl, many of which were eaten (though swans were a Royal prerogative) and from wild geese (such as St. Werburg put on parole—with the regrettable result that her cook made some of the virtuous birds into a pie).

The difference between scalded feathers and 'dry pulled' is known to any domesticated woman, but the difference aroused much indignation against users of the former in the wardens of the fellowship of the craft of the upholsterers in the reign of Henry VII). Clean feathers, plucked dry from a freshly killed fowl, need no more dressing than a good bake in the oven, after bread, but feathers once soiled require soaking and boiling and cleaning, and to mix the two qualities was indeed wrongful. Sometimes to-day (so probably also in Henry VII's time) if the feathers will not pull easily, the countrywoman puts the bird, head down, in a pail of warm water, which makes plucking less like a snowstorm, but is not too good for the feathers.

Some upholstery was stuffed with horse-hair. Good high-grade curled hair is now the most expensive and highly commended filling, but the medieval filling was probably 'refuse hair', soaked off the hides by the tanners.

Neats' hair, deers', and goats' hair were also used, but a frequent complaint against that hair which is wrought in lime vats, points to a process continued to-day, for the pelts are to-day wrought in lime, which reduces them to bare hides, clean snow-white fat, and a mass of soaked-off hairs (see p. 183). The fat goes to the wholesale pastry cooks, the trimmings to jelly squares and glue merchants. The hair, usually neats' hair, treated in this way was only supposed to be used in building construction as an ingredient in fibrous plaster, and it was probably not the country sellers' fault if the town buyer used it for pillows.

Fen down (which was accused of breeding worms) was almost certainly reed mace, for quantities of reed were cut for thatching work. Its use as vegetable filling was of old defended thus :

" Only in the first year that it (fen down) was gathered it would breed a worm while it was green, but being a year old, the stuff was as sweet and clean as any feathers. And for this, order was taken by them that none should be used until it were a year old, gathered by the poor people in the Isle of Ely. For the mingling of this stuff with feathers or any other stuff to deceive any was a thing they never heard of or knew."

Other vegetable fillings have been used, at various times, for various purposes such as bog cotton (*Eriphorum polystachym*), sphagnum moss, still prepared as filling and used for surgical cases, oat chaff, and straw which is described in the earlier section on straw.

The following is a list of plants that were used before the importation of Kapock down, though now few (except the sphagnum) are used by country people.

Feather top grass (*Calamagrostis* Epigejos), Moor Silk (*Polytrichum commune*), Bog Moss (*Genus Sphagnum, Sphagnum cvmbefolium, Sphagnum cuspidatum, Sphagnum aculifolium, Sphagnum squarrosum*), Stag's Horn (*Lycopodium clavatum*) (for saddlery *only*, we believe), Down Thistle (*Onopordon acanthium*), Cotton Grass or Bog Cotton (*Eriophorum polystachym capitatum*), and Reed Mace from fenland reed.

The old feather beds were made in the country, but I have never found full size mattresses being made in England by country people. (As a nation, our bedding is deplorably bad.) But I have found small ' beds ' made for babies' cots, filled with oat chaff. And many a district nurse would welcome their general adoption. The texture is healthy and porous, they are easy to make, light to shake, and when soiled, the filling can be changed and renewed instantly. I have failed to trace any ' tradition ' or locate any ' district ' to this bedding. They are made and used here and there, by rich and poor alike. Apparently, any ' family ' who ' know to it ', use it. It seemed slightly more common in Yorkshire, so perhaps the oat chaff owes something to the German influence in that district, and the plenitude of oat chaff in the North. Having been raised on one of these oat chaff mattresses myself, I can testify that they are soft enough for comfort and firm enough for health.

Incidentally, the old-fashioned crochet and knitted wool shawls loosely doubled and folded that were used with these non-conducting, porous, chaff mattresses combined the qualities of the most scientific modern production of aerated rubber and cellular wove blanketing. So that the most old-fashioned bed, made in England, is still actually the most sanitary and up-to-date.

The straw palliasse and straw bed are described under ' Straw ' on p. 74.

Sphagnum moss is also gathered and made up into surgical dressings, being very absorbent and having a high iodine content, is also a natural antiseptic.

§ 9. *Feathers*

The usage and making-up of feathers now only belongs to some isolated farms and old-fashioned country houses. Few country people now collect them for pillows and beds. A feather bed never wears out, and up to the end of the last century they were in common use in the country. With the changing fashion, the huge feather beds that billowed in the cavernous four-posters were given away to retainers, or sold for a song at the breaking up of the large estates—so that now, in small cottages, and tiny country inns, you are sometimes given feather beds made of the finest white goose down and goose feathers.

Such beds, taking 6s. a lb. as an average price for white down, would cost by weight of feathers anything up to £25—and yet the feather bed may have been given away as ' unsaleable ', or the filling used up for pillows. A *good* feather pillow in a town store costs 10s., yet in cottages where the entire income would be about 30s. a week, there are sometimes huge pillows whose value would be £1 or 30s. each.

The ' feather ' period in the country dates from the days

of mixed farming, when the woman's perquisite was the poultry yard. She would sell the poultry 'dressed' (i.e. undressed) in the market and as she plucked she collected and sorted the feathers automatically. Once her own household was well stocked, she sold the surplus feathers.

Nowadays, the poultry expert and dealer between them account for the poultry in bulk, and the feathers are disposed of direct to the feather merchant, who, in turn, supplies the bedding and upholstery manufacturers, and they sell again to the furnishing retailers. So that several profits are laid on the feathers before the farmer's daughter buys them back, encased in artificial silk, on the instalment plan.

The few country people who still collect the feathers do it like this: the bird is 'cleared', that is, all its foul feathers are taken away, and then it is plucked on a newspaper or cloth. The feathers are then pinned up into a paper bag, and either put into the bread oven, when it is cooling down, or hung up near the fireplace, till they have 'won', that is, dried out completely. When there are bags enough they are cautiously emptied into the new 'tick' (which has been well soaped internally to prevent the feathers working through), and sewn up. The new pillow is then well beaten and shaken, and left in a warm place for an hour or so to 'fluff up', after which it is ready. Old recipe books give instructions for cleaning feathers in lime water, but I have never found any country woman who did it; they wash them in bags, held down with clean stones in a clear stream.

White goose feathers were considered the best; goose-down, which they usually called 'dawn', was sold for a good price for making trimmings, puffs, and babies' pillows. Swans-down, too, was valuable. In the country where geese and white ducks are bred, white feathers are most prized, but common mixed feathers are good for ordinary use.

You must never use 'quill' feathers, for the quills will stick through the cloth. And—it is said—feathers from birds of flight destroy sleep. 'They put pigeon feathers in my pillow,' says Catherine Linton in *Wuthering Heights*, 'no wonder I could not die.'

Goose quills are sold to manufacturers for making paint

brushes and fishing floats, but this is a commercial enterprise between the poulterers and the manufacturers. The only things that are made from goose quills by the country people now, are home-made fishing floats and teats for bottle-fed lambs.

In country houses, goose-wings are much prized by the lady's maid and the housemaid, for nothing so well takes the last spot of dust from velvet as a damp goose wing ; and the country girl still keeps one in her housemaid's box for banisters and corners. In farmhouses I have seen them nailed to a strip of firewood and used for brooms for the hearth, or to clear out the white ash from a bake-oven.

NOTES

Sea Coast and Riverside

Boats of all kinds are still built by small craftsmen, though this craft is changing fast : coracles, punts and rafts are made on some rivers : ship tackle, canvas gear, weather canvas, sails, nets and fish-nets for deep sea and river, also seine-nets : floats for nets, of wood, cork, metal or glass (often blown in a town but mounted in the country) : some rope and rope goods, twine and string, including special cords such as Monsil ties for nets : wood and bone netting pins, back-steyn for walking on pebble beaches, and mudboards for the mud flats.

From the beaches, shells are made into a form of lime, and some seaweed burnt to give kelp and other products : of the common white seaweed variously called Dorset weed, carrageen, or Irish moss, nothing but jellies, soups, puddings, and cough mixtures are made in the country, but it is gathered for the use of leather workers, bookbinders, printers, painters, dyers, sweet makers, and others : flints gathered on pebble beaches are only used locally for building, but the black 'glass' flints are sold to make English flint glass.

Domestic

Some twenty different varieties of cheese were originally made by country women (most of them now reduced to a commercial mediocrity) : five or six varieties of hams, and various 'makes' of bacon : pies of all kinds, including the famous Melton Mowbray pork pies : sauces such as Worcester sauce and Yorkshire relish : smoked, dried, pressed and potted meats, especially near ports, for shipping stores, guaranteed to last for months on the old sailing boats : much anti-scorbutic and special ointments and medicines for sea folk (broom-water—extracts of broom and gorse flowers—forms a basis for many of these ; it was used for sheep in medieval times, before tar was known) : various animal and veterinary medicines.

Sand and Gravel Pits

Sand and gravel pits employ many country people and many old mine dumps are worked as small industries. The spar waste from the mines in Derbyshire is used for rough-casting and in Flintshire the fine white quartz from an old lead mine is graded, washed, and sold.

INDEX